Introduction to Research Ethics and Academic Integrity

Amani Khalil Abu-Shaheen
Muaawia Ahmed Hamza • Sumayyia Marar
Editors

Introduction to Research Ethics and Academic Integrity

Springer

Editors
Amani Khalil Abu-Shaheen
Scientific Writing and Biostatistics Department
King Fahad Medical City, Riyadh Second Health Cluster
Riyadh, Saudi Arabia

Muaawia Ahmed Hamza
Biomedical Research Administration
King Fahad Medical City, Riyadh Second Health Cluster
Riyadh, Saudi Arabia

Sumayyia Marar
Grant Management Department
King Fahad Medical City, Riyadh Second Health Cluster
Riyadh, Saudi Arabia

ISBN 978-981-96-9677-2 ISBN 978-981-96-9678-9 (eBook)
https://doi.org/10.1007/978-981-96-9678-9

This Springer imprint is published by the registered company Springer Nature Singapore Pte Ltd.
The registered company address is: 152 Beach Road, #21-01/04 Gateway East, Singapore 189721, Singapore

Preface

The idea of writing an edited book titled *Introduction to Academic and Research Integrity* is an initiative from the editors as an attempt to highlight the importance of academic and research integrity and to address the recent challenges. The editors were provoked by the lack of a unified strategy to deal with the use of generative artificial intelligence tools (GenAI) in scientific writing. Having the blueprint ready for this book, we started to invite scholars from the academic integrity community. Thanks to the authors of the chapters who agreed to contribute to this work and showed a high level of commitment in abiding by the blueprint and deadline.

The book is written in very simple language with a logical progression that familiarizes readers from different disciplines with its contents. The beauty of this book is that the authors are from different academic and cultural backgrounds. It starts with a pain-free introduction and offers simple definitions for the most common terminology used in the subsequent chapters. The introductory authors address several important aspects in the area of academic and research integrity, including the most common misconduct that breaches integrity; the recent dilemma on the use of GenAI tools in scientific writing; policies and their importance in maintaining academic integrity; and some scenarios for cases of breaching academic integrity and how to tackle them.

The chapters of this book are intended to broaden the horizons of the readers in the area of academic and research integrity. In addition to that it opens their eyes and encourages them to conduct research studies into this very fertile area and to develop new strategies to cope with any challenges that emerge in the future.

The editors would like to thank the authors of the chapters and the members of the entire academic integrity community for their dedication and enthusiasm in spreading the culture of academic integrity. Their hard work and enthusiasm were the real inspiration to continue.

Also, we would like to thank Dr. Zakia Saleem and Dr. Khadijah Mukhtar for their untiring efforts in helping us with a literature review of Chap. 3. Additionally, we appreciate the invaluable technical assistance provided by Basma Abdelazeem Abdalla and Bashayer Badawi Saeed.

Riyadh, Saudi Arabia — Amani Khalil Abu-Shaheen
Riyadh, Saudi Arabia — Muaawia Ahmed Hamza
Riyadh, Saudi Arabia — Sumayyia Marar

Declaration

The authors declares that using ChatGPT, an AI language model developed by OpenAI, to refine the language and enhance the readability of this book. After using this tool, the authors reviewed and edited the content as needed and took full responsibility for the chapter's content.

Contents

Introduction to Research and Academic Integrity

1

Muaawia Ahmed Hamza

This opening chapter aims to introduce readers to key terms and definitions, providing a foundation to make their journey through the book smooth and enjoyable. The topic at hand is particularly timely, given the rapid rise of artificial intelligence (AI) tools, which have sparked significant division within academic and research communities. On one side, some view AI as a valuable asset, offering new opportunities for learning and productivity, yet also grappling with the ethical responsibility of using these tools without undermining students' ability to develop their own writing skills. On the other side, some strongly oppose the use of AI in academic settings, raising concerns about its impact on integrity and authenticity. This growing debate has become a focal point for scholars and researchers committed to preserving academic research integrity.

1.1 What Is Research?

Having an extensive search into the literature, we found that there is a wider consensus on the definition of research. It is defined as an active systematic process of gathering, documenting, analyzing, and interpreting information in order to produce new knowledge (Kumar 2022; Hiebert et al. 2023). Nevertheless, it can be conducted to expand existing knowledge or to confirm or challenge the findings of previous research studies. It follows a robust methodology that is most appropriate for not only achieving the study's goals but also ensuring that the protocol used is replicable for other researchers (Kumar 2022). One of the key characteristics of scientific research is its emphasis on accuracy at every stage, from formulating the

M. A. Hamza (✉)
Biomedical Research Administration, King Fahad Medical City, Riyadh Second Health Cluster, Riyadh, Saudi Arabia

A. K. Abu-Shaheen et al. (eds.), *Introduction to Research Ethics and Academic Integrity*, https://doi.org/10.1007/978-981-96-9678-9_1

research question to reporting findings and sharing them with the scientific community. It is a laborious resource-consuming process. Scholars overcome these difficulties through collaboration and cooperation with different institutions and grant-funding agencies.

1.2 Why Is research Important?

Scientific research plays a critical role in addressing societal challenges. As mentioned earlier, the primary goal of research is to generate new knowledge. Its importance extends beyond building capabilities and promoting further research; it is also essential for fostering knowledge-based economies and driving economic growth. In many countries, prioritizing research and increasing public investment in it have become key indicators of national prosperity (UNESCO 2021; Confraria et al. 2023). Many funding agencies and endowments worldwide now allocate resources to specific research themes, reflecting this priority. Additionally, prestigious prizes are awarded annually to recognize and encourage scholars who conduct impactful, transformative research.

1.3 Different Types of Research

Scientific research can be either descriptive, exploratory, or explanatory (Kumar 2022). Descriptive research is conducted to describe, explain, and validate the findings of a study. It focuses on answering one research question through data collection and analysis. That can be achieved with very simple tools and at low cost, i.e., questionnaires or interviews (Creswell 2015). Exploratory research (from its name) is meant to navigate possible answers to a question through a systematic approach. Findings of exploratory research studies usually lead to more specific research questions. Last is the explanatory research or causal research. This type of research is conducted to test the extent of the relationship between the cause and effect by investigating several variables. These categories of research are also differentiated by the types of study design, sample size, sampling techniques, and data collection tools.

1.4 The Research Process

The research process involves several interconnected steps, where changes in any step require a review of the subsequent ones (Creswell 2015). It begins with identifying the research problem, followed by a thorough search of existing literature related to the problem to gain a deeper understanding of the subject. The literature review must be conducted carefully and critically, selecting reliable sources with a high degree of credibility (Galvan 2017). The knowledge gained through the literature informs the third step, which focuses on refining the research questions. This step, known as question formulation or statement of purpose, involves articulating the research question clearly and objectively, ensuring it is feasible within a given time frame. Once established, the research must select the most appropriate study design. The study design serves as the blueprint of how the research will be conducted to reach the expected outcomes. It should specify how data will be collected, documented, and which tools will be used for analysis. All steps in this phase must be presented in a logical sequence. Initiating data collection and analysis without a well-defined design will lead to failure, preventing the form meeting its intended objectives.

Data analysis is a crucial step in the research process and includes both sample size calculation and sampling technique. Instrument design and data collection tools have to be prepared to collect reliable data. To do so, one has to make sure that the instrument has been validated using the suitable validity approach. Also, it should be reliable and applicable, and the data generated is generalizable (Kumar 2022).

Using the right tools for data analysis produces meaningful results. The data analysis tools should not only present the results but also provide insights into the possible explanations and underlying causes. Researchers have to be careful when it comes to the selection of the data analysis tool, as using the wrong one can lead to results that contradict the research hypothesis, potentially confusing those attempting to follow the methodology. The choice of analysis tool depends on the nature of the data collected and the specified research question being addressed.

Finally, writing the research report is the concluding step in the research process. It has to reflect the underlying results. The report must be written in an academic writing style, characterized by clarity of the language used, critical analysis, and logical flow. Clear reports make it easy for the readers to follow and increase its impact. Furthermore, references both in the body and conclusion must be consistent (Fig. 1.1).

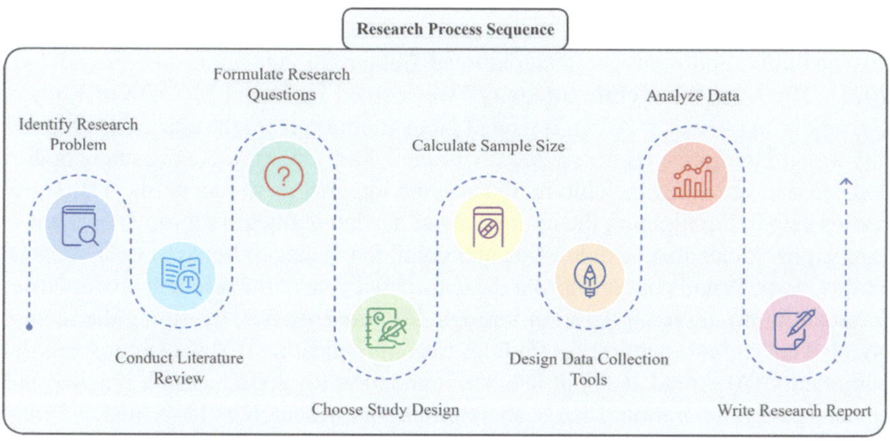

Fig. 1.1 AI-generated drawing illustrates the steps itemized in the research process

1.5 Research Ethics

Research ethics entails adhering to established standards of conduct in all stages of the research process as mentioned earlier (Sutrop et al. 2020). The key elements to be considered by researchers include confidentiality or respecting the participants' autonomy; informed consent; no harm to those who agreed to be part of the study subjects; and honesty and justice (Resnik 2020). Chapter 2 discusses these key ethical elements in more detail. All members of the research and academic communities must be made aware of these key elements (Resnik 2020). Part of the scholars' responsibility is to obtain ethical approval to conduct their studies before they start. There are different terms used for the committees granting ethical approvals. In some countries, they are known as the Institutional Review Boards (IRBs) or Research Ethics Committees (RECs) (Mehta et al. 2023). These committees' review is not constrained to make sure that everything involved in a given study is necessary, safe, and appropriate (Mehta et al. 2023). Also, they ensure that the principal researcher and the rest of the research team members are qualified to conduct the study. The IRBs and RECs may not approve a study if the storage of biospecimens and data generated are not explicitly mentioned. The role of these committees extends to monitoring the progress of the research studies at the execution stage and reviewing the final reports before disseminating them to the scientific community (Mehta et al. 2023). The IRBs and RECs mostly are independent committees to avoid bias and conflict of interest. They usually have a free hand to impose sanctions on those who encounter research misconduct. The presence of ethical committees helps in decreasing research misconduct and acts as quality assurance to ensure research integrity (Mehta et al. 2023).

1.6 Academic and Research Integrity

Academic integrity is best defined by the International Center for Academic Integrity (ICAI) "as a commitment to six fundamental values: honesty, trust, fairness, respect, responsibility, and courage" (International Center for Academic Integrity [ICAI] 2021). The term **"academic integrity"** was coined by Donal McCabe of Rutgers University in (Aronson 2024). It gained great attention from the academic community around the globe (Bertram 2008; Chesney 2009). Now, there are many bodies dedicated to spreading the culture of academic integrity as shown in Fig. 1.2. Those bodies help in highlighting the importance of academic integrity through awareness campaigns. It can also be defined as upholding the values of honesty, trust, responsibility, respect, and courage during the entire life cycle of the research process from formulating the research question through designing studies, executing them, analyzing data, and interpreting the findings to communicating it to the larger scientific community (Al-Amad 2017). It lays the foundation for good research practice and encourages collaboration. Like academic integrity, researchers have to show great commitment to the fundamental values mentioned in the Singapore Statement on Research Integrity, 2010 (Singapore Statement on Research Integrity 2010).

Fig. 1.2 Institutions dedicated to spreading the culture of academic integrity around the world

Maintaining research and academic integrity is the responsibility of everyone in the scholastic community (Al-Amad 2017). There is line of evidence that suggest that strong relationship between adopting integrity in education context and integrity in the future workplace (Harding et al. 2004; Guerrero-Dib et al. 2020). It was also found that the culture of academic and research integrity passes through the generations in one campus. Though early scholars attempted to explore the reasons behind some misconduct, evidence remains limited and scattered. However, the past couple of decades witnessed the birth of many institutions dedicated to spreading the culture of integrity in academic and research settings. These institutions organized awareness activities, conferences, workshops, and summer schools and established many specialized journals and periodicals. Some of these institutions include ICAI in the USA, the European Network for Academic Integrity (ENAI), the Australasian Academic Integrity Network (AAIN), the Asian Association for Academic Integrity (AAAI), the African Research Integrity Network (ARIN), the International Institute for Research and Action on Academic Fraud and Plagiarism (IRAFPA), and the Committee on Publication Ethics (COPE).

In the Middle East North African (MENA) region, the first Asia-Middle East-Africa Conference on Academic and Research Integrity (ACARI) was held in 2023 in Dubai, UAE. There are many ongoing intercontinental collaborative research projects to investigate some phenomena and to spread integrity culture, e.g., Facing Academic Integrity Threats (FAITH) project, Partnership on University Plagiarism Prevention (PUPP), and many other multicenter studies. Moreover, they help individual institutes in developing their integrity policies and building their capabilities. It is also worth mentioning some of the pioneers who dedicated their time to focus on the area of academic integrity. Among those scholars are Don

McCabe of Rutgers University, USA, the founder of ICAI 1992 who coined the academic integrity term; Teddi Fishman, USA, former director for the ICAI, the late Tracey Bretag of University of South Australia who left a great legacy of dedication to "academic integrity research"; Martine Peters, the director of PUPP; Sarah Elaine Eaton, University of Calgary, Canada; Irene Glendenning, Coventry University, UK; Salim Razi, Canakkale University, Turkie; Özgür Çelik, Balikesir University, Turkey; Zeenath Reza Khan, Wollongong University, Dubai, UAE; Tomáš Foltýnek, co-founder and president of the board of ENAI since establishment up-to-date; Sonja Bjelobaba, Uppsala University, Sweden; Laura Ribeiro, Porto, Portugal; Daniele Fanelli, UK; Robin Crockett, UK; Mary Davis, UK; Thomas Lancaster, Imperial College London, UK, who coined the term "contract cheating"; Shiva Sivasubramaniam, University of Roehampton, London, UK; Debora Weber-Wulff, Berlin, Germany; Oliver Trevisiol, University of Konstanz, Germany; Milan Ojstersek, University of Maribor, Slovenia; and Michelle Bergadaà, University of Geneva, Switzerland.

1.6.1 Plagiarism

In 2000, the US federal government set a uniform definition for research misconduct as fabrication, falsification, or plagiarism (FFP) (Resnik et al. 2015). Plagiarism, which implies stealing others' ideas or part of the published work without giving credit to the original authors, is the most ancient and prominent form of academic and research misconduct. Chapter 3 further elaborates on this form of misconduct. It will give a historical background into its evolution, types, detection methods, and how to prevent it. Although attention has now been focused on the threats imposed by emergence of the AI text-generating tools, plagiarism remains a big challenge. In the past two years following the advent of the ChatGPT, the challenge of AI text-generating tools has put the entire academic community under great pressure on how to keep it under control. Post-plagiarism a new term has been coined to describe the contemporary AI era.

1.6.2 Responsible Scholarship

Academic scholars are expected to be responsible multitaskers (Rana 2024). Their role is not confined to teaching and producing knowledge through scientific research, but they are also expected to contribute to the larger community. Any member of the academic scholarly community would want to be a responsible scholar. Academics usually have their individual goals of catching up with scholar impact matrices, i.e., citation counts, publishing, and personal growth, which supposedly goes hand in hand with their institutional goals of obtaining accreditations and reserving a high ranking among other academic entities (Rana 2024). A responsible scholar is expected to be equipped with several characteristics including demonstrating professionalism, applying critical thinking, embracing lifelong learning, proactively engaging in scholarly community activities as a reviewer and editor, training young scholars as a mentor, and contributing to building own institution's reputation (Rana 2024).

1.7 The Public Good of Research

The benefits of research are vast and cannot be captured in a single paragraph; however, some of the most visible advantages that impact daily lives will be highlighted. All the recent discoveries that have shaped and improved the quality of contemporary life are the result of extensive research. Without the dedication of scholars from different disciplines, society would not have achieved its current advancements. For instance, all the earth-breaking discoveries in the medical field significantly enhanced the quality of life for patients with chronic diseases, enabled the development of safe vaccines against communicable diseases, and facilitated early diagnosis of complex diseases through personalized medicine. The revolution in engineering and information technology which transformed our way of living at lightning speed resulted from extensive research studies. The research outcomes led to high-yield crops and livestock production, making life and population growth sustainable.

1.7.1 Importance of Research Integrity

The vast scholarly achievements that shape today's societies are fruits of extensive and diligent research studies. All scholarly work collectively contributes to the ever expanding body of knowledge. Hence, maintaining the research integrity encourages governments and the public sector to invest in research and continuously provide support (Yarborough and Sharp 2002).

1.8 Consequences of Research Misconduct

Trust, transparency, openness to critique, and collaboration are key elements in a healthy scholarly community (National Academies of Sciences, Engineering, and Medicine 2017). Failure to maintain any of those elements will result in distortion of the reputation of individual scholars, research groups, and their institution. Intentionally reporting false information can lead to a significant waste of time and precious resources for attempting to replicate the findings using the same methods. Additionally, it undermines trust and respect from the wider public and scientific community (Yarborough and Sharp 2002).Moreover, funding bodies, whether governmental or private, are likely to become highly skeptical of proposals from the institutions associated with research misconducts, making it more difficult for those institutions to secure funding. Recent studies suggest that there is a strong relationship between academic misconduct and the country's corruption level (Orosz et al. 2018).

1.9 Role of Students, Academics, and Research Institutions in Promoting Research Integrity

Promotion of integrity in the academic (Boehm et al. 2009) and research ecosystem is a multidimensional process (Valkenburg et al. 2021). It is a shared responsibility of students, faculty, and staff. Academic integrity can be promoted by educating students and raising their awareness about the different forms of misconducts and malpractices. Encouraging and rewarding those who abide and adopt it in their actions and deeds can perpetuate academic integrity. On the other hand, keeping communications up-to-date in the academic and research ecosystem on emerging threats to academic integrity will assure vigilance and increased ethical sensitivity. Faculty and staff ought to encourage their students to contribute and participate in educational activities, campaigns, and conferences on academic integrity. All stakeholders of academic institutions should report any type of academic misconduct without hesitation to protect their own institutions. Individual institutions should develop their academic integrity policies and have them available on their websites. Updated and robust academic integrity policies safeguard individual researchers, their research groups, and institutions from the malpractices and ensure commitment to good research practices (Glendinning 2023).

References

Al-Amad S. Responsible conduct of research. In: Silvernam H, editor. Research ethics in the Arab Region. New York: Springer; 2017. p. 359–71.

Aronson JK. When I use a word … Academic integrity—a brief history, a proposed definition, guidelines, and violations. BMJ. 2024;387:q2303. https://doi.org/10.1136/bmj.q2303.

Bertram T. Academic integrity in the twenty-first century. In: A teaching and learning imperative. ASHE Higher Education Report; 2008.

Boehm PJ, Justice M, Weeks S. Promoting academic integrity in higher education. The Community College Enterprise. 2009;15:45–61.

Chesney T. Review of the book Academic Integrity in the Twenty-First Century: A Teaching and Learning Imperative. Rev High Educ. 2009;32(4):544–5. https://doi.org/10.1353/rhe.0.0088.

Confraria H, Ciarli T, Noyons E. Countries' research priorities in relation to the Sustainable Development Goals; 2023. https://doi.org/10.5281/zenodo.7823881.

Creswell JW. Educational research: Planning, conducting, and evaluating quantitative and qualitative research. 5th ed. Boston: Pearson; 2015.

Galvan JL. Writing literature reviews: a guide for students of the social and behavioral sciences. 7th ed; 2017.

Glendinning I. Developing and implementing policies for academic integrity – Management of Change. In: Curtis GJ, editor. Academic Integrity in the Social Sciences. Ethics and Integrity in Educational Contexts, vol. 6. Cham: Springer; 2023. https://doi.org/10.1007/978-3-031-43292-7_6.

Guerrero-Dib JG, Portales L, Heredia-Escorza Y. Impact of academic integrity on workplace ethical behaviour. Int J Educ Integr. 2020;16:2. https://doi.org/10.1007/s40979-020-0051-3.

Harding TS, Passow HJ, Carpenter DD, Finelli CJ. An examination of the relationship between academic dishonesty and professional behavior. In: Antennas Propagation Magazine, IEEE; 2004. p. 46. https://doi.org/10.1109/MAP.2004.1388860.

Hiebert J, Cai J, Hwang S, Morris AK, Hohensee C. Doing research: A new researcher's guide. Springer; 2023. https://doi.org/10.1007/978-3-031-19078-0.

International Center for Academic Integrity [ICAI]. The fundamental values of academic integrity. 3rd ed; 2021. www.academicintegrity.org/the-fundamental-valuesof-academic-integrity.

Kumar R. Research methodology: A step-by-step guide for beginners. 5th ed. London: SAGE; 2022.

Mehta P, Zimba O, Gasparyan AY, Seiil B, Yessirkepov M. Ethics Committees: Structure, Roles, and Issues. J Korean Med Sci. 2023;38(25):e198. https://doi.org/10.3346/jkms.2023.38.e198.

National Academies of Sciences, Engineering, and Medicine. Fostering Integrity in Research. Washington, DC: The National Academies Press; 2017. https://doi.org/10.17226/21896.

Orosz G, Tóth-Király I, Bőthe B, Paskuj B, Berkics M, Fülöp M. Roland-Lévy C Linking cheating in school and corruption. Revue Européenne de Psychologie Appliquée. 2018;68(2):89–97. https://doi.org/10.1016/j.erap.2018.02.001.

Rana S. Understanding roles and practices of a responsible scholar. FIIB Business Review. 2024;13(4):387–90. https://doi.org/10.1177/23197145241247869.

Resnik DB What is ethics in research & why is it important? National Institute of Environmental Health Sciences 2020. https://www.niehs.nih.gov/research/resources/bioethics/whatis/index.cfm.

Resnik DB, Neal T, Raymond A, Kissling GE. Research misconduct definitions adopted by U.S. research institutions. Account Res. 2015;22(1):14–21. https://doi.org/10.1080/08989621.2014.891943.

Singapore Statement on Research Integrity 2nd World Conference on Research Integrity. 2010. Singapore, 21–21 July. Available at: https://www.jsps.go.jp/english/e-kousei/data/singapore_statement_EN.pdf.

Sutrop M, Parder ML, Juurik M. Research ethics codes and guidelines. In: Iphofen R, editor. Handbook of Research Ethics and Scientific Integrity. Cham: Springer; 2020. https://doi.org/10.1007/978-3-319-76040-7_2-1.

UNESCO. UNESCO Science Report 2021. Paris: UNESCO; 2021.

Valkenburg G, Dix G, Tijdink J, de Rijcke S. Expanding research integrity: a cultural-practice perspective. Sci Eng Ethics. 2021;27(1):10. https://doi.org/10.1007/s11948-021-00291-z.

Yarborough M, Sharp RR. Restoring and preserving trust in biomedical research. Acad Med. 2002;77(1):8–14.

Research Ethics and Ethical Behavior in Research

2

Gökhan Hınız

Research ethics is integral to the integrity and credibility of scientific inquiry. This chapter aims to establish a comprehensive ethical framework that guides researchers in protecting participant rights and ensuring responsible research practices. The chapter defines key ethical principles such as respect for persons, beneficence, non-maleficence, and justice; elaborates on the significance of informed consent and confidentiality; and addresses ethical challenges in authorship, conflicts of interest, and research involving vulnerable populations. The chapter highlights the potentially severe consequences of ethical breaches, ranging from participant harm to the erosion of public trust. It also offers practical recommendations and updated guidelines in light of technological advancements and digital research environments. Important insights underscore that adherence to ethical standards not only fulfils legal obligations but also enhances the quality and societal impact of research. Ultimately, this chapter serves as a vital resource for researchers, aiming to empower them with the knowledge needed to conduct research with integrity and social responsibility. Research ethics is "*the principles, values, and standards that guide researchers to conduct their work in a morally responsible, trustworthy, and reliable manner.*"(Iphofen, 2020, p. 4). It encompasses a wide range of considerations: respect for persons, beneficence to research participants, justice, trust, informed consent, confidentiality, privacy, data management, violations, fraud, and the ethical treatment of both human and animal subjects and avoidance of plagiarism (Allen, 1997; Armond et al., 2021; Iphofen, 2020; Kaufman & Ramarao, 2005; Mallia, 2018; Sachs, 2011; Serpico, 2024).

Research ethics is crucial for preserving the integrity and credibility of scientific inquiry. Adherence to ethical standards ensures that the research process is conducted transparently and that the findings are reliable and trustworthy (Bruton & Sacco,

G. Hınız (✉)
School of Foreign Languages, Department of Foreign Languages, Çanakkale Onsekiz Mart University, Çanakkale, Türkiye

A. K. Abu-Shaheen et al. (eds.), *Introduction to Research Ethics and Academic Integrity*, https://doi.org/10.1007/978-981-96-9678-9_2

2018). It is both a moral duty and a legal necessity, enforced by many institutions and regulatory authorities. Some of the key ethical guidelines include the (United States Government, 1949; World Medical Association, 2013), the European Union (EU) Directive 2001/20/EC on Good Clinical Practice, the Charter of Fundamental Rights of the EU, and the European Convention on Human Rights (Pierre, 2018). Researchers are expected to work according to ethical norms, codes, and guidelines as well as their employers' and government's policies and regulations. This adherence forms the basis of research ethics and responsible conduct of research, as outlined by the organizations, institutions, and relevant authorities (Shamoo & Resnik, 2009; Steneck, 2006; Steneck & Bulger, 2007; Xie et al., 2024).

The importance of research ethics lies in its critical role in safeguarding participants from harm while ensuring that their rights, dignity, and welfare are respected and protected throughout the research process (Andanda & Wathuta, 2017; Kaufman & Ramarao, 2005; Lothen-Kline et al., 2003). Ethical guidelines and principles serve as frameworks to ensure that research is conducted with the highest standards of integrity and honesty, which are essential for preserving the trust and credibility of the research community (Andanda & Wathuta, 2017; Williams & Pigeot, 2016).

2.1 Key Principles of Research Ethics

Research ethics is guided by key principles that protect participants and ensure the integrity of the research process. These include respect for persons, beneficence and non-maleficence, justice, informed consent, data privacy, confidentiality, authorship ethics, conflicts of interest, and special protections for vulnerable populations.

2.2 Respect for Persons

Respect for persons emphasizes the significance of acknowledging and respecting the autonomy of research participants (Kraft et al., 2023; Buccini et al., 2009). This principle is most evident in the process of obtaining informed consent, where participants are given comprehensive information about the research and are allowed to make a voluntary decision about their participation (Buccini et al., 2009; Cascio & Racine, 2018). This ensures that participants are not coerced into participating in research and retain control over their involvement. Respecting participants involves honoring their autonomy and considering their needs, values, and experiences throughout the research process (Kraft et al., 2023).

Cascio and Racine (2018) extend the concept of respect in person-oriented research ethics to include recognition of participants' holistic personhood, valuing their capabilities and potential contributions, even when decision-making capacity is impaired. Respect for persons includes providing extra safeguards for vulnerable individuals such as children, to prevent them from being exploited in research (Sherwood & Parsons, 2021).

2.3 Beneficence and Non-maleficence

Beneficence and non-maleficence require researchers to maximize the benefits of their research while minimizing potential harm to participants (Kapp, 2006; O'Donoghue, 2023). Beneficence involves promotion of participants' well-being and ensuring that the research enhances its positive impact (De Roubaix, 2011). It involves designing studies to ensure that the potential benefits outweigh any risks, with measures in place to minimize those risks (Serpico, 2024).

Non-maleficence is the ethical obligation to avoid causing harm, urging researchers to meticulously assess and mitigate any potential risks associated with their studies (Lothen-Kline et al., 2003). This principle requires researchers to conduct thorough risk assessments, considering the potential impact of their research on participants. It is crucial for researchers to evaluate potential risks to participants' physical and psychological well-being and to implement strategies to prevent such risks.

Even in seemingly harmless studies, unforeseen consequences may arise, so risk assessments should involve multiple researchers and stakeholders to ensure thorough evaluation and minimize potential harm. Additionally, this principle is closely linked to confidentiality, such as when a participant might critique their employers, policymakers, or politicians during data collection on a seemingly low-risk topic could lead to unforeseen consequences if the identity is disclosed.

2.4 Justice

Justice in research ethics ensures that specific groups are neither unfairly burdened nor excluded from the potential benefits of research (Gostin & Powers, 2006), especially vulnerable populations who are at a higher risk of exploitation (Roffee & Waling, 2017). Justice also ensures that the selection of research subjects is equitable and that outcomes are accessible to those who may benefit the most (Dauda & Dierickx, 2017; MacKay & Saylor, 2020; Serpico, 2024). This principle also involves addressing disparities and ensuring that all participants are treated with fairness and respect.

2.5 Informed Consent in Research

The concept of informed consent is closely tied to the principle of respect for persons and highlights the importance of acknowledging and safeguarding individuals' autonomy. By obtaining informed consent, researchers equip participants with all the necessary information about the research—such as its purpose, potential risks, and benefits—before they decide whether or not to take part (Buccini, 2009; Connor & Owen, 2018; Davies et al., 2024; Marshall, 2006). This is especially important in studies involving vulnerable populations, where additional protections may be required to prevent exploitation or harm (Sherwood & Parsons, 2021). It respects participants' autonomy by providing the information needed for an informed decision.

The informed consent process is not only a chance to build rapport and establish clear expectations but also a crucial ethical practice to maintain trust between researchers and participants and guarantee that participants are fully informed about the nature of the research, the associated risks, and their right to withdraw from the study at any time (Davies et al., 2024; Kraft et al., 2023; Ligtenberg et al., 2024; Petrova et al., 2016; Sankary et al., 2023). Moreover, informed consent also contributes to the ethical conduct of research and fosters a research environment that is transparent, responsible, and just (Kaufman & Ramarao, 2005).

Information about the nature of the research must be communicated in a manner that is understandable to the participants, considering their socioeconomic, cultural, and educational backgrounds (Halkoaho et al., 2016). Obtaining consent from vulnerable populations, such as children, individuals with cognitive impairments, or those from marginalized communities, requires a contextual approach to vulnerability (Gordon, 2020) and carefully drafted informed consent using simplified language and with the involvement of gatekeepers or legal representatives if necessary to facilitate understanding (Sherwood & Parsons, 2021).

Informed consent is an ongoing process that continues until the research study is completed (Lentz et al., 2016; Kadam, 2017). Researchers should provide participants with updated information as the research progresses, particularly if new risks or benefits emerge that could affect their willingness to continue participating (Lentz et al., 2016). This continuous communication ensures that participants remain fully informed and able to make decisions about their involvement in the study at any point (Sherwood & Parsons, 2021). The challenges of obtaining informed consent are further amplified in the context of digital and online research, where traditional methods of consent may not be feasible (Rocha-Silva et al., 2023). In studies involving social media data, participants may not be fully aware that their data is being collected and analyzed, raising significant ethical concerns about consent and privacy (Rossi et al., 2022). While data from platforms like X (formerly known as Twitter) or publicly available internet blogs where users openly share their identities is public, ethical considerations still require consent. The public availability of data does not necessarily imply that individuals have consented to use their information for research (Fiesler & Proferes, 2018). In such cases, researchers should consider the privacy expectations that users might have despite the public nature of the platform. Navigating these challenges may involve anonymizing data to protect individual identities, using only aggregated data, or obtaining consent retrospectively if possible. Additionally, researchers should adhere to platform-specific guidelines and ethical codes that govern the use of publicly shared data. Researchers must develop innovative consent mechanisms that respect participants' autonomy while addressing the unique challenges posed by digital environments (Rocha-Silva et al., 2023).

Williams and Pigeot (2016) argued that while informed consent is essential, the traditional approach may be too rigid, often resulting in overly detailed consent forms that participants may not fully understand. They propose a more flexible approach that emphasizes trust and transparency rather than exhaustive detail.

- Build trust by ensuring that research processes are transparent and that institutions are accountable.
- Promote transparency by making research practices, procedures, and outcomes open to scrutiny.
- Share research information broadly, allowing various stakeholders to access and scrutinise it, contributing to overall transparency.
- Allow participants ongoing opportunities to consent to different aspects of the research over time, provided this is done in a manageable way.
- Situate consent and confidentiality within a broader framework of research accountability, ensuring they support the integrity of the research process while respecting participants' rights.

A consent form should clearly outline the key aspects of the research in accessible language, considering the participants' socioeconomic, cultural, and educational backgrounds. Researchers should also consider using visual aids or simplified language. Based on various guidelines, Sutrop and Lõuk (2020, p. 222) recommended that informed consent forms should include the following elements:

- Details about the voluntary nature of participation.
- The right to withdraw without consequences.
- Information on who is conducting and financing the study.
- Potential conflicts of interest.
- Study aim, reasons for the subject's inclusion.
- Risks and benefits.
- Procedures and time commitment.
- Compensation details.
- Handling of incidental findings.
- Researcher contact information.
- Processing of personal data (including what data is collected, whether special categories are included, how the data is processed, and who has access to it).

In digital environments, researchers can use dynamic consent models that allow participants to update their consent preferences as they learn more about the research. Online platforms can facilitate interactive consent processes, such as video explanations or digital forms with hyperlinks for additional information. Keeping the participants fully informed in a digital context might also involve providing regular updates and reminders about their participation and ongoing consent. This approach not only upholds ethical standards but also fosters trust and transparency, which are essential for the successful and responsible conduct of research.

2.6 Confidentiality

Researchers have a fundamental duty to maintain the confidentiality of participants' information, ensuring that their personal data is not disclosed without their explicit consent (Baez, 2002; Bos, 2020; Kaiser, 2009; Morse & Coulehan, 2014; Williams & Pigeot, 2016). It builds trust with participants and is essential for ensuring the reliability and validity of research findings (Bos, 2020; Giordano et al., 2007). Confidentiality is a legal mandate in many jurisdictions (Voigt & Von Dem Bussche, 2017). Regulations such as the General Data Protection Regulation impose strict controls on the handling and sharing of personal data (Sharkas et al., 2024). Confidentiality is closely linked to the ethical principles of respect for persons and non-maleficence, which highlight the importance of protecting individuals' privacy and ensuring that their participation in research does not expose them to harm (Kaiser, 2009). It involves safeguarding participants' identities and the data they provide, particularly in studies involving sensitive topics or vulnerable populations (Bos, 2020). Even when data is shared publicly such as in digital media, ethical guidelines stress the importance of considering the context and potential consequences to protect individuals' privacy and dignity (Rossi et al., 2022).

Giordano et al. (2007) argued that while confidentiality protects participants from stigma, liability, and social repercussions, allowing participants to waive confidentiality introduces potential ethical dilemmas: risks to third-party confidentiality, potential biases against those maintaining confidentiality, and the possibility of participants distorting information. Researchers must carefully address these issues during the consent process to balance participants' autonomy with the protection of others' confidentiality.

Bos (2020) distinguishes between anonymity, which conceals participants' identities to prevent researcher linkage, and confidentiality, which protects identifiable information from unauthorized disclosure. Anonymity provides a higher level of privacy by preventing identification altogether, while confidentiality enables researchers to collect and protect sensitive data, preserving participants' trust (Bos, 2020). Both are essential in safeguarding participants' privacy and ensuring ethical research practices, but the choice between them depends on the study's context and objectives.

In qualitative research, maintaining confidentiality can be particularly challenging due to the depth of personal information collected and the close relationships that often develop between researchers and participants (Morse & Coulehan, 2014). Researchers must maintain the anonymity of data and avoid collecting unnecessary identifying information to reduce the risk of breaches (Baez, 2002; Morse & Coulehan, 2014; Petrova et al., 2016). In small or tightly-knit communities, where individuals may be easily identifiable even with anonymized data, additional precautions are necessary to protect participant confidentiality (Damianakis & Woodford, 2012).

Confidentiality also extends to the management and storage of data, where researchers keep the data securely stored and only accessible to authorized people (McKibbin, 2021; Rossi et al., 2022). This is particularly important in the context of digital research, where data breaches and unauthorized access pose significant risks to participants' privacy. Researchers must establish robust data security protocols, including encryption and secure storage solutions, to safeguard research participants' information (Rafiq et al., 2022). Breaches of confidentiality can lead to significant consequences for participants, including social stigma, psychological harm, or legal repercussions (Giordano et al., 2007; Surmiak, 2020; Turcotte-Tremblay & Sween-Cadieux, 2018). For researchers, breaches of confidentiality can lead to loss of trust, damage to professional reputation, and legal liabilities (Kaiser, 2009). These risks highlight the need for strict confidentiality protocols and transparency with participants regarding how their data will be protected.

Confidentiality is a core principle in social research ethics, yet researchers encounter significant ethical dilemmas when they discover illegal activities or harm during their studies (Surmiak, 2020). Maintaining confidentiality must also take into consideration the duty to report serious harm or risks to participants or others (Kaiser, 2009). In some cases, researchers may face dilemmas where maintaining confidentiality could lead to harm, such as in studies involving illegal activities or severe psychological distress (Giordano et al., 2007) Confidentiality can be breached in cases where participants disclose suicidal thoughts (Lothen-Kline et al., 2003). Researchers need to make context-specific decisions rather than relying on rigid ethical rules, advocating for a more flexible and nuanced approach to handling confidentiality in research (Surmiak, 2020). In these situations, researchers must carefully weigh the potential harms and benefits of breaching confidentiality and seek guidance from colleagues, ethics review boards, or other ethical frameworks.

In conclusion, maintaining confidentiality in research is a critical ethical obligation that underpins the trust between researchers and participants. Researchers must take several precautions, including anonymizing data to remove any identifying information about participants, third parties, and the research setting, such as school or company names. Anonymization protects identification of individuals, thereby reducing the risk of unintended disclosure of sensitive information. This practice is especially important in qualitative research, where even small details can lead to the identification of participants in tightly knit communities or specific contexts. The use of software for data storage, along with locally stored data, requires encryption, password protection, and immediate anonymization to protect the confidentiality and prevent unauthorized access or data breaches.

For participants, breaches of confidentiality can lead to social stigma, psychological harm, legal repercussions, and other adverse outcomes. For researchers, such breaches can result in loss of trust, damage to professional reputation, and legal liabilities. These risks highlight the importance of strict adherence to confidentiality protocols and transparent communication between both parties. Confidentiality should not hinder the sharing of data necessary for scientific validation and progress; instead, it should be seen as a principle that enables responsible data sharing under strict safeguards rather than as a barrier to data sharing (Williams & Pigeot, 2016).

2.7 Authorship and Publication

Authorship and publication ethics ensure that credit is accurately attributed and that the scientific record reflects honest and transparent research practices. Authorship involves substantial contributions to the research, drafting or revising the manuscript, final approval of the version to be published, and accountability for the work (Aliukonis et al., 2020).

The two most common forms of unethical authorship are honorary (including individuals who did not significantly contribute) and ghost authorship (excluding deserving contributors) (Aliukonis et al., 2020). Honorary authorship includes gift, guest, and coercive authorship, where individuals are credited without substantial contribution. Ghost authorship involves significant contributors not being listed as authors. Both practices are widespread, with studies showing high rates of authorship misuse globally, especially in regions with less developed academic practices (Aliukonis et al., 2020). Such practices often stem from power dynamics, favoritism, and the pressure to succeed in a competitive academic environment (Smith et al., 2020). Another form of unethical authorship is contract cheating, "where students contract out their coursework to writers or workers, usually found via the internet, in order to submit the purchased assignments as their own work."(Walker & Townley, 2012, p. 27) Lancaster and Clarke (2011) first defined this academic dishonesty practice as the external outsourcing of academic tasks, with the work typically being original and challenging to detect using traditional plagiarism detection tools. Unlike traditional plagiarism, which involves copying existing content, contract cheating involves original work tailored to a student's specific assignment, making it more deceptive and harder to trace (Lancaster & Clarke, 2008). The increasing accessibility of information and services online has amplified the issue and therefore requires innovative approaches to combat this growing problem (Lancaster & Clarke, 2008; Walker & Townley, 2012) .

The criteria for authorship are frequently determined by international guidelines, such as those established by the International Committee of Medical Journal Editors (ICMJE). Adhering to these criteria helps prevent issues like gift and ghost authorship. According to the ICMJE's (ICMJE | Recommendations | Defining the role of authors and contributors, 2024) recommendations on defining the role of authors and contributors, authors must meet the following criteria:

- Substantial contributions to the conception or design of the work or the acquisition, analyze, or interpretation of data for the work and drafting the work or reviewing it critically for important intellectual content and final approval of the version to be published and agreement to be accountable for all aspects of the work in ensuring that questions related to the accuracy or integrity of any part of the work are appropriately investigated and resolved. Transparency in authorship is essential to ensure that credit is given where it is due and that the contributions of all team members are appropriately recognized. Transparency also involves disclosing potential conflicts of interest that could influence the interpretation of research findings.

- Substantial contributions to the conception or design of the work; or the acquisition, analysis, or interpretation of data for the work; and
- Drafting the work or reviewing it critically for important intellectual content; and
- Final approval of the version to be published; and
- Agreement to be accountable for all aspects of the work in ensuring that questions related to the accuracy or integrity of any part of the work are appropriately investigated and resolved.

Transparency in authorship is essential to ensure that credit is given where it is due and that the contributions of all team members are appropriately recognised regarding contribution, authorship and acknowledgements (McNutt et al., 2018; Smith et al., 2020). Transparency also involves disclosing potential conflicts of interest that could influence the interpretation of research findings (McNutt et al., 2018). Failure to manage authorship appropriately can lead to disputes, damage collaborations, and undermine the trust in the published work (Smith et al., 2020).

Douglas et al. (2022) examined various methods for determining the order of authors in academic publications and explored the ethical implications. The study highlighted that alphabetical order is commonly used due to its straightforward nature but may undervalue the contributions of individual researchers, leading to lower perceptions of distributive justice. In contrast, ordering authors by their contribution is perceived as a fairer method since it aligns author positions with their actual input. However, it can cause conflict if not clearly communicated. The study also discussed a hybrid approach that combines elements of both alphabetical and contribution-based ordering, which offers a potential compromise but still necessitates careful management of procedural fairness. It is important to establish clear and transparent authorship policies to address power dynamics within research teams and ensure ethical practices in assigning authorship orders (Douglas et al., 2022).

2.8 Conflicts of Interest

Conflicts of interest occur when researchers have personal, financial, or other interests that could potentially influence their work, leading to biased outcomes. It can negatively impact the integrity of the research process, as it may create situations where researchers' decisions and findings are swayed by external factors rather than objective evidence (Bruton & Sacco, 2018; Resnik, 2023). Addressing these conflicts transparently is essential to preserve the credibility of research. When researchers disclose and manage potential conflicts, they help safeguard the objectivity and reliability of their work, reinforcing public trust in the research process (Resnik, 2023). Transparency in handling conflicts of interest is crucial for maintaining the accountability of the research community to the public, as it reassures stakeholders that the research is conducted with integrity and free from undue influence (Bruton & Sacco, 2018; Friedman, 2002). By openly addressing these issues, researchers contribute to a culture of ethical responsibility, which is fundamental to the advancement of knowledge and the credibility of scientific inquiry (Anderson et al., 2013).

To ensure ethical disclosure of conflicts of interest, several practices can be adopted within the research community. One approach is the inclusion of disclosures in publications and presentations. For example, a researcher funded by a pharmaceutical company for a study should explicitly state this in related publications. An appropriate disclosure might read, "This research was funded by XYZ Pharmaceuticals. The authors have no other relevant affiliations or financial involvement with any organization or entity with a financial interest in the subject matter discussed in the manuscript. (AI-generated by ChatGPT-4o on 31.07.2024, based on similar publisher statements)." This level of transparency allows peers and readers to assess the potential for bias and make informed judgments about the findings (Resnik, 2023).

2.9 Considerations for Research Involving Vulnerable Populations

Research involving vulnerable populations requires heightened ethical sensitivity and stringent safeguards to ensure protection from harm and exploitation (Gordon, 2020). Vulnerable populations include individuals with limited capacity to give informed consent, such as children, people with cognitive impairments, and those from marginalized communities (Shivayogi, 2013). These groups are often subject to power imbalances that can influence their participation in research, necessitating additional ethical considerations. Researchers working with vulnerable populations, such as justice-involved youth, must navigate complex ethical considerations, including the need to protect the autonomy and confidentiality of participants (Quinn, 2015). Ethical reflexivity, or the continuous reflection on the ethical aspects of the research process, is crucial for addressing power dynamics and ensuring that research is conducted ethically (von Unger, 2021).

Conducting research with vulnerable populations, especially those from marginalized or indigenous communities, requires a deep understanding of cultural contexts and a commitment to respect participants' values and traditions (Tauri, 2018). Researchers need to collaborate with community leaders and stakeholders to build trust and ensure that the research is carried out in a way that is culturally sensitive and respectful (Amundsen & Msoroka, 2021). Engaging with the communities involved in research is essential for conducting research ethically and sharing the benefits (Tauri, 2018). This collaborative approach addresses ethical concerns and enhances the research's relevance and impact (Tauri, 2018).

2.10 Consequences of Unethical Research

Unethical research has often resulted in significant harm to participants, leading to lifelong psychological or physical disabilities (Algahtani et al., 2018). The lasting impact of these studies, such as the psychological trauma from the Monster Study and the physical harm in the Nazi experiments, highlights the severe consequences of neglecting ethical standards (Algahtani et al., 2018).

Unethical research has profound implications that extend beyond the immediate harm to participants and affect the broader scientific community and society. Ethical standards are essential not only for protecting individuals from physical injury, psychological distress, and social stigma but also for ensuring the integrity and credibility of scientific inquiry. Once breached, the repercussions are severe, undermining the validity, reliability, and reproducibility of research findings (Sivasubramaniam et al., 2021). Historical examples, such as the Tuskegee Syphilis Study, clearly illustrate how unethical practices can cause profound, long-lasting harm to individuals and communities, leading to a loss of trust in the scientific community (Alsan et al., 2019). This erosion of trust can diminish public confidence in scientific findings, particularly among marginalized groups who may feel exploited by past research practices. Additionally, unethical research often produces unreliable or invalid results, distorting the scientific record and leading to potentially harmful policy or clinical decisions. The consequences include legal repercussions for researchers and institutions, such as fines, sanctions, loss of professional licenses, as well as damage to reputation. These guidelines should emphasize transparency, accountability, and the protection of participants' rights, ensuring that the advancement of technology in research continues to align with the fundamental principles of research ethics.

2.11 Conclusion

In conclusion, research ethics serve as the cornerstone of responsible scientific inquiry, ensuring the protection of participants, the integrity of the research process, and the credibility of scientific findings. The principles outlined in this chapter underscore the critical importance of ethical considerations across all stages of research, from the protection of vulnerable populations to the responsible handling of data and the transparent reporting of methodologies. Adherence to these ethical standards is not merely a legal requirement but a moral imperative that upholds the trust between researchers and the public. As technology and artificial intelligence (AI) become increasingly integral to the research process, there is a growing need for clear guidelines on their ethical and transparent use. The deployment of AI in research design, data analysis, and even the automation of research tasks introduces new ethical challenges, including bias, privacy, and informed consent. Without proper guidance, using these technologies could compromise the ethical standards vital to the integrity of research. Therefore, developing and implementing comprehensive guidelines that address the ethical use of technology and AI in research practices is imperative. These guidelines should emphasise transparency, accountability, and the protection of participants' rights, ensuring that the advancement of technology in research continues to align with the fundamental principles of research ethics.

Acknowledgment During the preparation of this work, the author used generative artificial intelligence tools (ChatGPT and Grammarly) to improve readability and language. After using this tool/service, the author reviewed and edited the content as needed and takes full responsibility for the content of the publication.

References

Algahtani H, Bajunaid M, Shirah B. Unethical human research in the field of neuroscience: a historical review. Neurol Sci. 2018;39(5):829–34. https://doi.org/10.1007/s10072-018-3245-1.

Aliukonis V, Poškutė M, Gefenas E. Perish or publish dilemma: challenges to responsible authorship. Medicina. 2020;56(3):123. https://doi.org/10.3390/medicina56030123.

Allen C. Spies like us: when sociologists deceive their subjects. Lingua Franca. 1997;7(9):31–9.

Alsan M, Wanamaker M, Hardeman RR. The Tuskegee study of untreated syphilis: a case study in peripheral trauma with implications for health professionals. J Gen Intern Med. 2019;35(1):322–5. https://doi.org/10.1007/s11606-019-05309-8.

Amundsen D, Msoroka M. Responsive ethics: navigating the fluid research space between HREC ethics, researcher ethics, and participant ethics. Educ Rev. 2021;73(5):563–79. https://doi.org/10.1080/00131911.2019.1655392.

Andanda P, Wathuta J. Human dignity as a basis for providing post-trial access to healthcare for research participants: a south African perspective. Med Health Care Phil. 2017;21(1):139–55. https://doi.org/10.1007/s11019-017-9782-1.

Anderson MS, Shaw MA, Steneck NH, Konkle E, Kamata T. Research integrity and misconduct in the academic profession. In: Higher education. Dordrecht: Springer; 2013. p. 217–61. https://doi.org/10.1007/978-94-007-5836-0_5.

Armond A, Gordijn B, Lewis J, Hosseini M, Bodnár J, Holm S, Kakuk P. A scoping review of the literature featuring research ethics and research integrity cases. BMC Med Ethics. 2021;22 https://doi.org/10.1186/s12910-021-00620-8.

Baez B. Confidentiality in qualitative research: reflections on secrets, power and agency. Qual Res. 2002;2(1):35–58. https://doi.org/10.1177/1468794102002001638.

Bos J. Confidentiality. In: Bos J, editor. Research ethics for students in the social sciences. Cham: Springer; 2020. p. 149–73. https://doi.org/10.1007/978-3-030-48415-6_7.

Bruton SV, Sacco DF. What's it to me? Self-interest and evaluations of financial conflicts of interest. Res Ethics. 2018;14(4):1–17. https://doi.org/10.1177/1747016117739940.

Buccini LD, Caputi P, Iverson D, Jones C. Toward a construct definition of informed consent comprehension. J Empirical Res Hum Res Ethics. 2009;4(1):17–23. https://doi.org/10.1525/jer.2009.4.1.17.

Cascio MA, Racine E. Person-oriented research ethics: integrating relational and everyday ethics in research. Account Res. 2018;25(3):170–97. https://doi.org/10.1080/08989621.2018.1442218.

Clarke R, Lancaster T. Establishing a systematic six-stage process for detecting contract cheating. In: 2007 2nd international conference on pervasive computing and applications. IEEE; 2007. p. 342–7.

Connor J, Copland S, Owen J. The infantilised researcher and research subject: ethics, consent and risk. Qual Res. 2018;18(4):400–15. https://doi.org/10.1177/1468794117730686.

Damianakis T, Woodford MR. Qualitative research with small connected communities. Qual Health Res. 2012;22(5):708–18. https://doi.org/10.1177/1049732311431444.

Dauda B, Dierickx K. Viewing benefit sharing in global health research through the lens of Aristotelian justice. J Med Ethics. 2017;43(6):417–21. https://doi.org/10.1136/medethics-2015-102858.

Davies H, Kolstoe SE, Lockett A. Reshaping consent so we might improve participant choice (III)—how is the research participant's understanding currently checked and how might we improve this process? Res Ethics. 2024;20(3):604–12. https://doi.org/10.1177/17470161241235910.

De Roubaix J. Beneficence, non-maleficence, distributive justice and respect for patient autonomy—reconcilable ends in aesthetic surgery? J Plast Reconstr Aesthet Surg. 2011;64(1):11–6. https://doi.org/10.1016/j.bjps.2010.03.034.

Douglas HM, Elliott KC, Settles IH, Montgomery GM, Davis T, Nadolsky L, Cheruvelil KS. Authorship climate: a new tool for studying ethical issues in authorship. Account Res. 2022;31(5):403–27. https://doi.org/10.1080/08989621.2022.2140587.

Fiesler C, Proferes N. "Participant" perceptions of twitter research ethics. Soc Media Soc. 2018;4(1):205630511876336. https://doi.org/10.1177/2056305118763366.

Friedman PJ. The impact of conflict of interest on trust in science. Sci Eng Ethics. 2002;8(3):413–20. https://doi.org/10.1007/s11948-002-0041-7.

Giordano J, O'Reilly M, Taylor H, Dogra N. Confidentiality and autonomy: the challenge(s) of offering researcher participants a choice of disclosing their identity. Qual Health Res. 2007;17(2):264–75.

Gordon BG. Vulnerability in research: basic ethical concepts and general approach to review. Ochsner J. 2020;20(1):34–8. https://doi.org/10.31486/toj.19.0079.

Gostin LO, Powers M. What does social justice require for the public's health? Public health ethics and policy imperatives. Health Aff. 2006;25(4):1053–60. https://doi.org/10.1377/hlthaff.25.4.1053.

Halkoaho A, Pietilä A, Ebbesen M, Karki S, Kangasniemi M. Cultural aspects related to informed consent in health research. Nurs Ethics. 2016;23(6):698–712. https://doi.org/10.1177/0969733015579312.

ICMJE | Recommendations | Defining the role of authors and contributors. (2024) Retrieved August 2, 2024, from https://www.icmje.org/recommendations/browse/roles-and-responsibilities/defining-the-role-of-authors-and-contributors.html

Iphofen R. An introduction to research ethics and scientific integrity. In: Iphofen R, editor. Handbook of research ethics and scientific integrity. Switzerland AG: Springer; 2020. p. 3–13. https://doi.org/10.1007/978-3-030-16759-2_62.

Kadam R. Informed consent process: a step further towards making it meaningful! Perspect Clin Res. 2017;8(3):107. https://doi.org/10.4103/picr.picr_147_16.

Kaiser K. Protecting respondent confidentiality in qualitative research. Qual Health Res. 2009;19(11):1632–41. https://doi.org/10.1177/1049732309350879.

Kapp MB. Ethical and legal issues in research involving human subjects: do you want a piece of me? SSRN Electron J. 2006; https://papers.ssrn.com/sol3/Delivery.cfm/SSRN_ID953259_code218808.pdf?abstractid=953259&mirid=4

Kaufman CE, Ramarao S. Community confidentiality, consent, and the individual research process: implications for demographic research. Popul Res Policy Rev. 2005;24(2):149–73. https://doi.org/10.1007/s11113-004-0329-9.

Kraft SA, Duenas DM, Shah SK. Patient priorities for fulfilling the principle of respect in research: findings from a modified Delphi study. BMC Med Ethics. 2023;24(73) https://doi.org/10.1186/s12910-023-00954-5.

Lancaster T, Clarke R. The phenomena of contract cheating. In: IGI global eBooks; 2008. p. 144–159. https://doi.org/10.4018/978-1-59904-801-7.ch010.

Lentz J, Kennett M, Perlmutter J, Forrest A. Paving the way to a more effective informed consent process: recommendations from the clinical trials transformation initiative. Contemp Clin Trials. 2016;49:65–9. https://doi.org/10.1016/j.cct.2016.06.005.

Ligtenberg W, Stolper M, Molewijk B. Ethics support for ethics support: the development of the confidentiality compass for dealing with moral challenges concerning (breaching) confidentiality in moral case deliberation. BMC Med Ethics. 2024;25(49) https://doi.org/10.1186/s12910-024-01039-7.

Lothen-Kline C, Howard DE, Hamburger EK, Worrell KD, Boekeloo BO. Truth and consequences: ethics, confidentiality, and disclosure in adolescent longitudinal prevention research. J Adolesc Health. 2003;33(5):385–94. https://doi.org/10.1016/S1054-139X(03)00184-8.

MacKay D, Saylor KW. Four faces of fair subject selection. Am J Bioeth. 2020;20(2):5–19. https://doi.org/10.1080/15265161.2019.1701731.

Marshall PA. Informed consent in international health research. J Empir Res Hum Res Ethics. 2006;1(1):25–41. https://doi.org/10.1525/jer.2006.1.1.25.

McKibbin KJ, Malin BA, & Clayton EW. (2021). Protecting research data of publicly revealing participants. Journal of Law and the Biosciences, 8(2):1–12. https://doi.org/10.1093/jlb/lsab028

McNutt MK, Bradford M, Drazen JM, Hanson B, Howard B, Jamieson KH, Kiermer V, Marcus E, Pope BK, Schekman R, Swaminathan S, Stang PJ, Verma IM. Transparency in authors' contributions and responsibilities to promote integrity in scientific publication. Proc Natl Acad Sci. 2018;115(11):2557–60. https://doi.org/10.1073/pnas.1715374115.

Morse JM, Coulehan J. Maintaining confidentiality in qualitative publications. Qual Health Res. 2014;25(2):151–2. https://doi.org/10.1177/1049732314563489.

O'Donoghue K. Learning analytics within higher education: autonomy, beneficence and non-maleficence. J Acad Ethics. 2023;21(1):125–37. https://doi.org/10.1007/s10805-021-09444-y.

Petrova E, Dewing J, Camilleri M. Confidentiality in participatory research: challenges from one study. Nurs Ethics. 2016;23(4):442–54.

Pierre M. WASP (write a scientific paper): ethics approval for a research study (1). Early Hum Dev. 2018;124:46–8. https://doi.org/10.1016/j.earlhumdev.2018.04.022.

Rafiq F, Awan MJ, Yasin A, Nobanee H, Zain AM, Bahaj SA. Privacy prevention of big data applications: a systematic literature review. SAGE Open. 2022;12(2):215824402210964. https://doi.org/10.1177/21582440221096445.

Resnik D. Disclosing and managing non-financial conflicts of interest in scientific publications. Res Ethics. 2023;19(2):121–38. https://doi.org/10.1177/17470161221148387.

Rocha-Silva T, Nogueira C, Rodrigues L. Passive data collection on Reddit: a practical approach. Res Ethics. 2023; https://doi.org/10.1177/17470161231210542.

Roffee JA, Waling A. Resolving ethical challenges when researching with minority and vulnerable populations: LGBTIQ victims of violence, harassment, and bullying. Res Ethics. 2017;13(1):4–22. https://doi.org/10.1177/1747016116658693.

Rossi A, Arenas MP, Kocyigit E, Hani M. Challenges of protecting confidentiality in social media data and their ethical import. In: 2022 IEEE European symposium on security and privacy workshops (EuroS&PW); 2022. https://doi.org/10.1109/eurospw55150.2022.00066.

Sachs B. Going from principles to rules in research ethics. Med Ethics eJournal. 2011; https://doi.org/10.1111/j.1467-8519.2009.01744.x.

Sankary LR, Zelinsky ME, Ford PJ, Blackstone EC, Fox RJ. Overcoming barriers to informed consent in neurological research: perspectives from a national survey. Res Ethics. 2023;19(1):42–61. https://doi.org/10.1177/17470161221131497.

Shamoo AE, Resnik DB. Responsible conduct of research. Oxford: Oxford University Press; 2009. https://doi.org/10.1093/acprof:oso/9780195368246.001.0001.

Sharkas GF, El-Masry R, Abdel-Ghany S, El Baz A, Abou-Elsaad T, Kassab AA, Badran DH, Bashir AY. Satisfaction of academic medical staff with integrated medical curriculum: an exploratory multinational survey. BMC Med Educ. 2024;24:1483. https://doi.org/10.1186/s12909-024-06468-x.

Serpico K. The Belmont report doesn't need reform, our moral imagination does. Research Ethics. 2024;20(3):559–73. https://doi.org/10.1177/17470161241235772.

Sherwood G, Parsons S. Negotiating the practicalities of informed consent in the field with children and young people: learning from social science researchers. Res Ethics. 2021;17(4):448–63. https://doi.org/10.1177/17470161211014941.

Shivayogi P. Vulnerable population and methods for their safeguard. Perspect Clin Res. 2013;4(1):53. https://doi.org/10.4103/2229-3485.106389.

Sivasubramaniam SD, Cosentino M, Ribeiro L, Marino F. Unethical practices within medical research and publication—an exploratory study. Int J Educ Integr. 2021;17(7) https://doi.org/10.1007/s40979-021-00072-y.

Smith E, Williams-Jones B, Master Z, Larivière V, Sugimoto CR, Paul-Hus A, Shi M, Diller E, Caudle K, Resnik DB. Researchers' perceptions of ethical authorship distribution in collaborative research teams. Sci Eng Ethics. 2020;26(4):1995–2022. https://doi.org/10.1007/s11948-019-00113-3.

Steneck NH. Fostering integrity in research: definitions, current knowledge, and future directions. Sci Eng Ethics. 2006;12(1):53–74. https://doi.org/10.1007/pl00022268.

Steneck NH, Bulger RE. The history, purpose, and future of instruction in the responsible conduct of research. Acad Med. 2007;82(9):829–34. https://doi.org/10.1097/acm.0b013e31812f7d4d.

Surmiak A. Should we maintain or break confidentiality? The choices made by social researchers in the context of law violation and harm. J Acad Ethics. 2020;18(3):229–47. https://doi.org/10.1007/s10805-019-09336-2.

Sutrop M, Lõuk K. Informed consent and ethical research. In: Iphofen R, editor. Handbook of research ethics and scientific integrity. Springer; 2020. p. 12–4. https://doi.org/10.1007/978-3-030-16759-2.

Tauri JM. Research ethics, informed consent and the disempowerment of first nation peoples. Res Ethics. 2018;14(3):1–14. https://doi.org/10.1177/1747016117739935.

Turcotte-Tremblay A, Sween-Cadieux EM. A reflection on the challenge of protecting confidentiality of participants while disseminating research results locally. BMC Med Ethics. 2018;19(S1). https://doi.org/10.1186/s12910-018-0279-0.

Quinn, C. R. (2015). General considerations for research with vulnerable populations: Ten lessons for success. Health & Justice, 3(1):1–7. https://doi.org/10.1186/s40352-014-0013-z

United States Government Trials of war criminals before the Nuremberg Military Tribunals under Control Council Law No. 10. U.S. Government Printing Office; 1949. Vol. 2, p. 181–2.

Voigt P, Von Dem Bussche A. The EU general data protection regulation (GDPR). Cham: Springer; 2017. https://doi.org/10.1007/978-3-319-57959-7.

Von Unger H. Ethical reflexivity as research practice. Hist Soc Res. 2021;46(2):186–204.

Walker M, Townley C. Contract cheating: a new challenge for academic honesty? J Acad Ethics. 2012;10(1):27–44. https://doi.org/10.1007/s10805-012-9150-y.

Williams G, Pigeot I. Consent and confidentiality in the light of recent demands for data sharing. Biom J. 2016;59(2):240–50. https://doi.org/10.1002/bimj.201500044.

World Medical Association. World Medical Association Declaration of Helsinki: Ethical principles for medical research involving human subjects. JAMA. 2013;310(20): 2191–94. https://doi.org/10.1001/jama.2013.281053.

Xie J, Stockley D, Hastings Truelove A, Marlin S, Zand R, Payne J, Miller M, Soleas E. Needs and preferences of REB members in the development of a new TCPS 2 training program in Canada. Res Ethics. 2024;20(3):613–31. https://doi.org/10.1177/17470161231218173.

Contravenes of Academic Integrity

3

Lubna Ansari Baig and Mahwish Arooj

The teaching, training, and practice of education and research is the most noble profession since the faculty is responsible for building future leaders. Academic integrity refers to the highest standards of professionalism, inclusive of honesty, ethical practices, fairness, trust, and respect for contextual values in the conduct of academic responsibilities (Hafizha, 2022; Miron et al., 2021). Violations of academic integrity have made headlines in the popular press which devalues the noble profession of education and scholarship in institutions of higher education. This chapter highlights the importance of academic integrity and its noncompliance in higher education institutions. It will also describe the types of academic misconduct and give suggestions to ensure that the highest level of academic integrity is maintained. The chapter will delve into all types of academic misconduct inclusive of plagiarism, fabrication, cheating, etc. Real-world examples will be given with references for different types of academic misconduct and recommendations for managing them.

3.1 What Is Academic Integrity?

Academic integrity refers to the moral principles that guide the creation and sharing of knowledge in academia, ensuring that students and staff members acquire skills and produce work in a fair, responsible and honest manner. The contraventions to academic integrity are common across all professions, but they are particularly concerning in education and healthcare practice. The academic misconduct during education often carries over into future healthcare practice, with both students and faculty sometimes unaware of the misconduct committed. The contraventions of

L. A. Baig (✉) · M. Arooj
Department of Medical Education, University College of Medicine and Dentistry, University of Lahore, Lahore, Pakistan
e-mail: mahwish@uol.edu.pk

A. K. Abu-Shaheen et al. (eds.), *Introduction to Research Ethics and Academic Integrity*, https://doi.org/10.1007/978-981-96-9678-9_3

academic misconduct include cheating, helping other students unfairly, copying from work of others without acknowledgment, sharing exam materials, and fabricating data, among others. These are misconducts and are considered punishable by the laws of the country and relevant licensing bodies. There is a growing issue of dishonesty in academia, hence maintaining academic integrity is essential for the credibility of higher education institutions. Boehm et al. (2009) mentioned that widespread availability of gadgets and technology has raised the level of academic dishonesty.

3.2 Academic Misconducts

Also known as breach to academic integrity, academic misconducts devalue both education and research and undermine the achievements of honest individuals since suspicion is raised for all. It is considered academic dishonesty and involves a range of unethical practices. Failing to report misconduct is also considered an academic misconduct. In a recent study by Azam and Naeem, trainees wrote fake histories and were not aware that it is academic misconduct. Generally, academic misconduct includes cheating, fabrication, collusion, and sabotage (Hughes & McCabe 2006; Henning et al. 2019; Perry 2010).

3.3 Fabrication

This is a very severe type of academic misconduct in which a researcher/faculty alters the data or manipulates the information without actual work. More specifically, fabrication is creation of false findings or outcomes, fictitious survey responses, or reports from tests that were never conducted (https://www.bmartin.cc/pubs/16hai/Vaux.html). Trust is undermined by fabrication, as it not only leads to false scientific conclusions but also compromises the authenticity of the research work (Martyn 2003). The fabrication may not only mislead other researchers and scholars; it can also pose risk to public safety due to wrong results (Badley 2009). One professor at a leading university in Alberta was charged with plagiarism where he fabricated data and published papers to get faculty tenure. His tenure was revoked, yet he continued as professor at the university. Writing fake histories in wards during postgraduate training is also another form of fabrication and considered serious academic misconduct.

3.4 Falsification

Falsification is the act of intentionally misrepresenting or misleading the data and information. It involves altering, manipulating, or omitting the existing data or information to support false findings or results, e.g., changing methods, materials, research instruments, images, results, etc. (Eungoo & Hwang 2020) Like other forms of academic misconduct, falsification undermines trust in research and damages the reputation of a noble profession (Eastman JK et al. 2008). In falsification, study results are manipulated, whereas in fabrication, the entire study is invented. This is a common form of academic misconduct, as researchers may manipulate results to secure publication, knowing that most journals favor significant findings. A professor and president of Howard University in the United States was fired after resigning as president as she was charged with falsifying her results and 50 other counts of plagiarism. https://www.quora.com/Claudine-Gay-the-antisemitic-racist-who-plagiarised-and-faked-all-her-researchhas-been-fired-from-Harvard-but-should-she-also-have-to-repay-her-salary-face-prison-time-and-also-be-banned-from-alluniversities-and.

3.5 Strategies to Overcome Fabrication and Falsification

To safeguard the integrity of academic contribution, researchers need to maintain rigorous professional standards and ensure rigor, accuracy, and honesty in their work. Policies need to be in place to ensure faculty follows guidelines in avoiding all forms of academic misconduct. There are statistical tools (chi-square test for uniformity of digit distributions) that can be used for detecting fabrication (https://blog.bjbms.org/academic-misconduct-in-research-fabrication-falsification-plagiarism-paper-mills/).

3.6 Collusion

This refers to helping another colleague in ways not allowed legally. Any collaboration between students done for gaining undue advantage for assignments, sharing answers in exams, conducting research projects, etc., is considered collusion. Faculty members collaborating on projects related to research, academics, and other academic work that is not allowed legally is all forms of collusion. Marking proxies for friends in undergraduate and postgraduate institutions is very common and it is not legal. A study by Hafeez K. et al. reported that almost 85% of medical college students agreed to mark proxies for their friends (Hafeez et al 2013).

Collusion can be formal with contracts done for joint work or informal with no official agreements. Below are some examples of collusion:

- Submitting an assignment done by another student as your own.
- Two or more students working on an assignment meant for individual work and submitting paraphrased versions.
- Marking proxies or asking friends to mark for them.
- Collaborating on research projects even if it is not in the relevant domain to increase publications. This is a very common practice to gain advantage for promotion, and some faculty are unaware it is considered as academic misconduct (Hughes & McCabe 2006).

3.7 Ghostwriting

Ghostwriting involves having someone else write your assignment, thesis, synopsis, or research paper for publication on your behalf. Ghostwriters who are anonymous are hired to write biographies and speeches for celebrities and political leaders and develop presentations for bosses in healthcare and business documents (Singh & Remenyi 2016). It is a common practice in many developing countries to ask qualified juniors to write manuscripts and position papers to get in favor of the supervisors/bosses. Unfortunately, in both developing and the developed world, this is not considered academic misconduct by students and faculty.

3.8 Cheating/Fraud

Cheating typically refers to copying during exams, plagiarizing from another student's assignment, or engaging in similar dishonest practices in assessments that determine grades or pass/fail outcomes (Burgason et al. 2019). This was more evident during COVID-19 as exams were held online and students had multiple means to cheat. With the massive growth in artificial intelligence models, cheating has become more common and more difficult to identify. In a blog from 2007 Alex Gillis, it was mentioned that the cheating trends in Canada and the United States have been rising in postgraduate students as well https://universityaffairs.ca/features/feature-article/cheating-themselves/. In one Canadian study involving 11 institutions, high numbers of graduate students cheating in exams and copying from other students' assignments have been documented. In online courses, 60% of the students agreed that they cheated, and doing so gave them a higher grade (Hughes & McCabe 2006). Cheating impacts individuals and institutions and a culture of fraud develops which can destroy personal and professional values. Nevertheless, fraud is more related to financial embezzlement which includes, but not limited to, filing wrong time sheets, marking proxies, and submitting false travels to gain financial benefits. It also includes inventing data when it's supposed to be collected in the field. In one scenario, a project director has been fired, and he requested that he may be allowed to resign from his position at a leading university in Karachi after he invented data and claimed honorarium for training he had never conducted.

3.9 Sabotage

It is referred to as deliberately preventing another student/faculty from completing their work. Sabotage may be physically destroying property of another student/faculty like computer, equipment, lab data, etc (Schmelkin et al. 2008). It also includes damaging another person's reputation by falsely accusing them of data fabrication, disapproving their work during peer review, etc. This is serious academic misconduct and, in most countries, punishable under law. The purpose of sabotage in students is generally to ensure that one gets a better grade than the other. In academics, the faculty may be competing for tenure positions or research, and by sabotaging the work of others, they get an undue advantage.

3.10 Measures to Promote Academic Integrity (De Maio & Dixon 2022)

3.10.1 Faculty Engagement

Faculty members are responsible for educating students about the importance of academic integrity and preventing misconduct. They must also communicate effectively with one another and report any incidents that breach academic integrity (Rahimian 2023). There is a dire need of training programs for faculty regarding policies for academic integrity and identification of misconduct (Gottardello & Karabag 2022).

3.10.2 Institutional Role

Institutions must establish clear educational policies and integrate academic integrity into their culture. They should provide faculty training and maintain strict vigilance to ensure the effective implementation of these policies. Collaboration with relevant national and international organizations is encouraged to keep up with the massive and rapid growth of artificial intelligence (Spain & Robles 2011).

3.10.3 Roles of Students in Promoting Academic Integrity

Students frequently engage in academic misconduct due to time constraints, stress, and insufficient understanding of academic misconduct. Institutions should implement educational programs to enhance awareness of academic integrity and offer necessary support. Engaging with students in their learning processes and fostering a supportive environment can decrease the chances of misconduct and promote integrity (Garza Mitchell & Parnther 2018). It is also the responsibility of the students to honor the code of conduct and help the institution in the development and promotion of academic integrity policies (Meizlish 2003).

3.11 Plagiarism

Plagiarism, one of the most common issues in academia, involves using others' ideas and words as one's own without proper citation (Helgesson & Eriksson 2018a). It is a significant threat to the integrity of academics as it compromises the credibility and reliability of scholarly work (Wiwanitkit et al. 2013). It can pose negative consequences, including diminishing the quality of research work and misleading future research.

Examples of plagiarism as reported in media

- A professor at a leading university in Edmonton, Alberta, was charged with plagiarism as he copied his address to graduates from an article in the "*New Yorker*"; he lost his position as dean but continued as a professor. https://macleans.ca/education/uniandcollege/when-professors-plagiarize/.

- Another professor at the University of Calgary Alberta was charged with plagiarism as the published paper had text from sources not duly referenced; the paper was retracted. https://macleans.ca/education/uniandcollege/when-professors-plagiarize/.
- A professor in the United States was demoted to associate and demoted to co-director after he was charged with plagiarism; he had copied directly from websites and other sources in his book without referencing. https://www.azcentral.com/story/news/local/tempe/2015/07/10/edu-popular-asu-professor-matthew-whitakerdemoted-plagiarism-incident/30000997/.

3.12 Types of Plagiarism

Each type of plagiarism undermines the ethical foundation of academic work and can distort scholarly contributions (Awasthi 2019). Self-plagiarism is reusing one's own previously published work without citation. It may seem harmless but is nevertheless considered unethical as it undermines intellectual honesty and can distort academic integrity (Bretag & Mahmud 2009).

3.13 Types of Self-Plagiarism

1. **Duplicate Publication:** Publication of the same research paper in multiple journals without informing the editor (Roig & Ph 2015).
2. **Text Recycling (Salami Slicing):** Reusing sections of previous work in new publications without proper citation. One example is the reuse of methods or literature reviews of previously published work in new research (Padmanabhan & Batra 2021).
3. **Redundant Publication:** Publishing multiple research articles with different titles but the content is similar with minor variations in data (Cicutto 2008).

In academia, the most common types of plagiarism as identified in the literature are listed in Table 3.1 (Shah et al. 2022).

3.14 Complicity

The facilitation of unethical behavior in academia and research is termed as complicity. It involves cheating and helping other students unlawfully, which can violate academic integrity. It emphasizes the concept that maintaining the integrity of academia or research is a collective effort of all the relevant members (Leathwood & Read 2013). The direct facilitation of misconduct manifests as a committed effort to allow cheating in exams and research.

Table 3.1 Types of plagiarism

Sr. no	Types of plagiarism	Characteristics	Severity
1	Direct	Completely copying text without citation	Serious
2	Self-plagiarism	Reusing your work without citation	Moderate
3	Source not valid or error 404	Using false citations	Moderate
4	Mosaic	Combining, mixing, and rephrasing one or more sources	Serious
5	Shotgun	Multiple submissions of the same manuscript to different journals	Serious
6	Colluding	Researchers cite each other's work with the intention of their promotion	Moderate
7	Secondary sources	Data used from another paper without reading the paper and/also without citing it	Moderate
8	Paraphrasing	Rephrasing words as own idea without citation	Serious
9	Verbatim	Use of quotation marks to copy ideas without citation	Serious
10	Stitching	No original work and content are copied and pasted with a citation from sources	Moderate
11	Slicing	Publishing multiple components of the same study instead of one publication	Serious
12	Picture manipulation	Copying the image to present as serious original work	Serious

3.15 Complicity in Research

Collaborative Misconduct Researchers may collaborate on projects without disclosing conflicts of interest, allowing unethical practices among co-authors, such as data fabrication and publication of falsified research (Helgesson & Eriksson 2018b).

Unethical Publishing The pressure to publish can lead to unethical publishing practices among researchers such as salami slicing or publishing without proper citations or acknowledgment (Wawer 2019).

3.16 Strategies to Overcome Plagiarism and Complicity

1. **Institutional policies:** In most countries, national policies have been developed that outline the consequences for those caught plagiarizing (Moin et al. 2020).
   (a) Example: Professors at leading universities in the United States https://www.azcentral.com/story/news/local/tempe/2015/07/10/edu-popular-asu-professor-matthew-whitaker-demotedplagiarism-incident/30000997/ and Pakistan https://www.dawn.com/news/amp/974140 were convicted and, as a result, got demoted and their tenures revoked.

2. **Plagiarism detection software** is being used in most higher education institutions locally and internationally.
3. **Familiarizing with the publication policies and adhering to the journal rules against plagiarism can help authors avoid self-plagiarism (Recommendations on Publication Ethics Policies for Medical Journals || WAME, n.d.) (Policies || WAME, n.d.).**
   (a) Teaching students and faculty about the consequences of plagiarism and mass education to disseminate information about the importance of academic integrity.
   (b) Institutions can initiate educational programs to raise awareness about self-plagiarism among students and faculty. (Guidelines on Ethical Conduct and Publication | Journal of University College of Medicine and Dentistry, n.d.), (Publication Ethics | Proceedings, n.d.).
4. **Guidelines to avoid self-plagiarism (Zimba & Gasparyan 2021):**
   (a) Citation: Cite previously published work when reusing it in a new publication.
   (b) Transparency with Editors: Inform editors at the time of submission about the overlaps of content with the previous publications.
   (c) Limit Reuse of Text: Rewrite new content to reflect fresh perspectives and limit the use of large sections from prior publications.
   (d) Ethical Collaboration: Ensure ethical collaboration with all the authors and acknowledge the contributors.

3.17 Conclusion

Academic integrity is the foundation of both education and research. Violations such as plagiarism, cheating, falsification, or fraud undermine the trust and credibility essential to academic institutions. It is the responsibility of the entire academic community to uphold and promote these ethical standards, ensuring that academic work remains a true reflection of knowledge, effort, and honest inquiry. Only through a shared commitment to integrity can the value of education and research be preserved, and higher education institutions must enforce policies that uphold the highest standards of academic integrity.

References

Awasthi S. Plagiarism and academic misconduct: a systematic review. DESIDOC J Libr Inf Technol. 2019;39(2):94–100. https://doi.org/10.14429/djlit.39.2.13622.

Azam M, Naeem SB. Academic integrity among medical students and postgraduate trainees in the teaching hospitals of South Punjab Pakistan. Health Inf Libr J. 2022;39(4):377–84. https://doi.org/10.1111/hir.12458.

Badley G. Academic writing as shaping and re-shaping. Teach High Educ. 2009;14(2):209–19. https://doi.org/10.1080/13562510902757294.

Boehm PJ, Justice M, Weeks S. Promoting academic integrity in higher education introduction and review of literature. Community Coll Enterp. 2009:45–61.

Bretag T, Mahmud S. Self-plagiarism or appropriate textual re-use? J Acad Ethics. 2009;7(3):193–205. https://doi.org/10.1007/s10805-009-9092-1.

Brown T, Isbel S, Logan A, Etherington J. Predictors of academic integrity in undergraduate and graduate-entry masters occupational therapy students. Hong Kong J Occup Ther. 2020;33(2):42–54. https://doi.org/10.1177/1569186120968035.

Burgason KA, Sefiha O, Briggs L. Cheating is in the eye of the beholder: an evolving understanding of academic misconduct. Innov High Educ. 2019;44(3):203–18. https://doi.org/10.1007/s10755-019-9457-3.

Cicutto L. Plagiarism*: avoiding the peril in scientific writing. Chest. 2008;133(2):579–81. https://doi.org/10.1378/chest.07-2326.

De Maio C, Dixon K. Promoting academic integrity in institutions of higher learning: what 30 years of research (1990–2020) in Australasia has taught us. J Coll Charact. 2022;23(1):6–20. https://doi.org/10.1080/2194587x.2021.2017972.

Eastman JK, Iyer R, Reisenwitz TH. The impact of unethical reasoning on different types of academic dishonesty: an exploratory study. J Coll Teach Learn. 2008;5(12):7–16. https://doi.org/10.19030/tlc.v5i12.1211.

Eungoo K, Hwang H-J. The consequences of data fabrication and falsification among researchers. J Res Publ Ethics. 2020;1(2):7–10.

Garza Mitchell RL, Parnther C. The shared responsibility for academic integrity education. New Directions Community Coll. 2018;2018(183):55–64. https://doi.org/10.1002/cc.20317.

Gottardello D, Karabag SF. Ideal and actual roles of university professors in academic integrity management: a comparative study. Stud High Educ. 2022;47(3):526–44. https://doi.org/10.1080/03075079.2020.1767051.

Guidelines on Ethical Conduct and Publication | Journal of University College of Medicine and Dentistry. (n.d.). Retrieved September 28, 2024, from https://journals.uol.edu.pk/jucmd/Guidelines-on-Ethical-Conduct-and-Publication

Hafeez K, Khan M L-u-Z, Jawaid M, Haroon S. Academic misconduct among students in medical colleges of Karachi, Pakistan. Pak J Med Sci. 2013;29(3):699–702. https://doi.org/10.12669/pjms.293.3300.

Hafizha R. Pentingnya Integritas Akademik. J Educ Couns. 2022;1(2):115–24. https://doi.org/10.32627/jeco.v1i2.56.

Helgesson G, Eriksson S. Plagiarism in research. In: Getting to good: research integrity in the biomedical sciences; 2018a. p. 262–73. https://doi.org/10.1007/s11019-014-9583-8.

Helgesson G, Eriksson S. Responsibility for scientific misconduct in collaborative papers. Med Health Care Philos. 2018b;21(3):423–30. https://doi.org/10.1007/s11019-017-9817-7.

Henning MA, Chen Y, Ram S, Malpas P. Describing the attributional nature of academic dishonesty. Med Sci Educ. 2019;29(2):577–81. https://doi.org/10.1007/s40670-019-00710-8.

Hughes JMC, McCabe DL. Understanding academic misconduct. Can J High Educ. 2006;36(1):49–63. https://doi.org/10.47678/cjhe.v36i1.183525.

Krueger L. Academic dishonesty among nursing students. J Nurs Educ. 2014;53(2):77–87. https://doi.org/10.3928/01484834-20140122-06.

Leathwood C, Read B. Research policy and academic performativity: compliance, contestation and complicity. Stud High Educ. 2013;38(8):1162–74. https://doi.org/10.1080/03075079.2013.833025.

Malik AA, Hassan M, Rizwan M, Mushtaque I, Lak TA, Hussain M. Impact of academic cheating and perceived online learning effectiveness on academic performance during the COVID-19 pandemic among Pakistani students. Front Psychol. 2023;14(March):1–7. https://doi.org/10.3389/fpsyg.2023.1124095.

Martyn C. Fabrication, falsification and plagiarism. QJM. 2003;96(4):243–4. https://doi.org/10.1093/qjmed/hcg036.

Meizlish D. Promoting academic integrity in the classroom. CRLT Occasional Papers. 2003;20:1–12. http://www.crlt.umich.edu/sites/default/files/resource_files/CRLT_no20.pdf

Miron J, Eaton SE, McBreairty L, Baig H. Academic integrity education across the Canadian higher education landscape. J Acad Ethics. 2021;19(4):441–54. https://doi.org/10.1007/s10805-021-09412-6.
Moin M, Dahar MA, Yousuf MI. Evaluative study of plagiarism in post graduate research in Punjab, Pakistan. Glob Soc Sci Rev. 2020;V(II):10–21. https://doi.org/10.31703/gssr.2020(v-ii).02.
Okoroafor AU, Henning MA, Chibuike OM, Rajput V. Disclosing academic dishonesty: perspectives from Nigerian and New Zealand health professional students. Ethics Behav. 2016;26(5):431–47. https://doi.org/10.1080/10508422.2015.1055494.
Padmanabhan S, Batra P. Publication ethics and malpractices: an insight. J Indian Orthodontic Soc. 2021;55(2):113–21. https://doi.org/10.1177/03015742211007968.
Perry B. Exploring academic misconduct: some insights into student behaviour. Act Learn High Educ. 2010;11(2):97–108. https://doi.org/10.1177/1469787410365657.
Policies || WAME. n.d. Retrieved September 28, 2024, from https://www.wame.org/policies
Publication Ethics | Proceedings. n.d.. Retrieved September 28, 2024, from https://proceedings-szmc.org.pk/index.php/szmc/publication-ethics
Rahimian M. Faculty engagement in addressing academic integrity challenges. Can Perspect Acad Integr. 2023;6(1):76915. https://doi.org/10.55016/ojs/cpai.v6i1.76915.
Recommendations on Publication Ethics Policies for Medical Journals || WAME. n.d.. Retrieved September 28, 2024, from https://wame.org/recommendations-on-publication-ethics-policies-for-medical-journals
Roig, M. (2015). Avoiding plagiarism, self-plagiarism, and other questionable writing practices: a guide to ethical writing. 1–71. https://ori.hhs.gov/sites/default/files/plagiarism.pdf
Satalkar P, Shaw D. Is failure to raise concerns about misconduct a breach of integrity? Researchers' reflections on reporting misconduct. Account Res. 2018;25(6):311–39. https://doi.org/10.1080/08989621.2018.1493577.
Schmelkin LP, Gilbert K, Spencer KJ, Pincus HS, Silva R. A multidimensional scaling of college students' perceptions of academic dishonesty. J High Educ. 2008;79(5):587–607. https://doi.org/10.1080/00221546.2008.11772118.
Shah JN, Shah J, Baral G, Baral R, Shah J. Types of plagiarism and how to avoid misconduct: pros and cons of plagiarism detection tools in research writing and publication. Nepal J Obstet Gynaecol. 2022;16(2):3–18. https://doi.org/10.3126/njog.v16i2.42085.
Singh S, Remenyi D. Plagiarism and ghostwriting: the rise in academic misconduct. S Afr J Sci. 2016;112(5–6):1–7. https://doi.org/10.17159/SAJS.2016/20150300.
Spain JW, Robles MM. Academic integrity policy: the journey. Bus Commun Q. 2011;74(2):151–9. https://doi.org/10.1177/1080569911404407/ASSET/1080569911404407.FP.PNG_V03.
Stander M. Strategies to help university students avoid plagiarism: a focus on translation as an intervention strategy. J Furth High Educ. 2020;44(2):156–69. https://doi.org/10.1080/0309877X.2018.1526260.
Wawer J. How to stop salami science: promotion of healthy trends in publishing behavior. Account Res. 2019;26(1):33–48. https://doi.org/10.1080/08989621.2018.1556099.
Wiwanitkit S, Wiwanitkit S, Wiwanitkit V. LETTER TO THE EDITOR. Attitude toward Plagiarism. 2013;17(3):88973667.
Zimba O, Gasparyan AY. Plagiarism detection and prevention: a primer for researchers. Reumatologia. 2021;59(3):132–7. https://doi.org/10.5114/reum.2021.105974.

Plagiarism Prevention

4

Saeed Ur Rehman

4.1 What Is Plagiarism?

Historically, the term "plagiarism" originates from the Latin word *plagiarius*, which signifies a "kidnapper" (Taylor 2024, p. 5). This etymological legacy gives an interesting meaning to plagiarism, which is the act of falsely claiming another person's intellectual work and then submitting it as one's own to a university or a publisher. The Roman poet Martial (AD 38–41–c.103) used the word *plagiarius* in his *Epigrams* to convey the sense that his verses had been kidnapped by Fidentinus, as if the literary theft was an intellectual abduction (Cresswell 2010, p. 329). Martial uses "kidnapping" and "stealing" of verses distinctly, though both expressions target the same plagiarist. The idea of kidnapping that is inherent in *plagiarius* is related to runaway slaves and literary theft is indicated by the Latin word *furtum* (Seo 2009).

In the sixteenth century, the word "plagiary," an English adaptation of the Latin *plagiarius*, appeared and conveyed the same notion of literary theft or abduction (Shaw 1982, p. 327). The time of its appearance is significant, as it coincides with the period following the invention of the printing press. Before the development of mass printing, plagiarism was largely a nonissue because, in the age of calligraphy, copying a scholarly treatise was not financially rewarding. Calligraphy was such a laborious art form that copying by hand was not considered plagiarism. However, with new technological developments enabling mass printing, it became possible for writers and publishers to reproduce books on a large scale and earn profits. This marked the first time that ideas in their written form could be sold as commodities, with books emerging as a form of capital.

S. U. Rehman (✉)
Comparative Humanities, Habib University, Karachi, Pakistan
e-mail: saeed.rehman@ahss.habib.edu.pk

A. K. Abu-Shaheen et al. (eds.), *Introduction to Research Ethics and Academic Integrity*, https://doi.org/10.1007/978-981-96-9678-9_4

This shift led to the development of copyright, which was originally called a "printing patent" and was granted by the sovereign (Mann 2010, p. 57). The royal permission to print a book eventually evolved into copyright, turning ideas into registered private property. By the mid-eighteenth century, copyright laws had clearly defined plagiarism, and public attitudes toward the practice underwent a significant transformation.

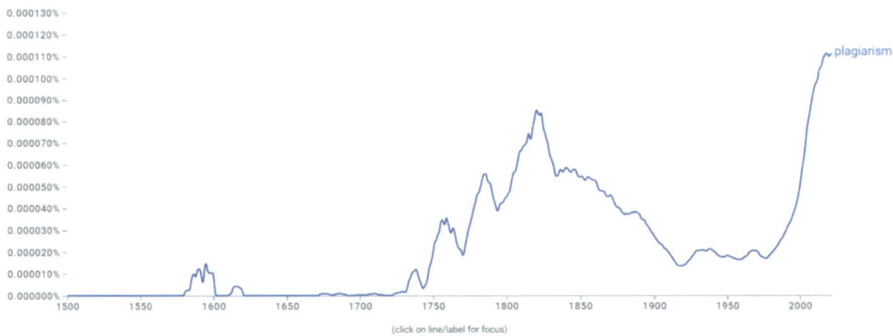

Source: Google Books Ngram Viewer (Google 2024)

Google's Ngram Viewer displays the usage of the term "plagiarism," which shows aggregated data on the occurrence of a term from 1500 to 2022. As the graph shows, the term appeared in English in the sixteenth century which marked the period when publishing houses were busy securing their exclusive rights to publish specific books and preventing their competitors from publishing the same books. Ideas were being transformed into private property for the first time at a large scale because of the technological shift in book production. This demonstrates that the history of copyright is deeply intertwined with the rise of capitalism.

Now that we have explored the historical context in which plagiarism emerged as a violation of copyright, we are better equipped to discuss the idea or the problem in its contemporary context. The present-day legal regimes and institutions are well established. In the contemporary setting, plagiarism is a major violation of ethical and professional standards.

Plagiarism is the act of implicitly or explicitly presenting somebody else's work or idea as one's own, regardless of whether it is done intentionally or unintentionally. It includes the use of someone else's published or unpublished work/material without acknowledging the source; the use of generative AI, writing bots, or similar technological aids to generate part or all of a scholarly work; submitting an old assignment or dissertation as new or submitting an already-published book as a new book; using part or all of a previously submitted or published piece of writing in a new form; submitting the same or similarly worded articles with different titles to multiple journals; or simply failing to cite one's sources correctly and diligently.

In the contemporary world, plagiarism is considered a major violation of academic and intellectual integrity that extends far beyond what used to be the simple act of copying (before the invention of the printing press). At its core, it constitutes a dishonest appropriation of the intellectual work, ideas, or expressions of others without respecting the right of the original creator to be recognized as the original creator. It means now ideas belong to persons in the same way that property does. Legally speaking, a person is someone who can hold the title of a property, an individual, or an incorporated legal entity. For example, the magazine *The Economist* does not list the names of individual authors under the titles of its published articles. However, this does not mean the articles can be freely copied or used without permission. Although there is no single human author explicitly identified as the copyright holder, the magazine, as an incorporated entity, holds the rights to the work. This demonstrates that copyright, like other forms of property, can belong to organizations, corporations, or other legal entities. Both individuals and entities, such as organizations and corporations, can act as authors and owners of copyright.

Plagiarism is an issue of academic integrity, competence, or due diligence. Detected in the work of a scholar, it reflects poorly on their character and undermines their reputation and casts a shadow on the rest of their work. Plagiarism may have consequences beyond this damage to reputation or academic standing. Duty bound to cultivate original thinking and research integrity, academic institutions strive to weed out plagiarism by imposing stringent disciplinary action against anyone who is found guilty of this violation. Copyright holders of a work that has been plagiarized also have recourse to civil and punitive action in most situations and are often willing to take that road.

For our purposes, plagiarism is the failure to properly cite or acknowledge the sources used in a piece of writing. This includes failing to cite at all, failing to cite fully, or failing to cite properly. It can result from sloppy work habits, carelessness, or unscrupulousness. For example, it may still be considered plagiarism even when one does everything else correctly but fails to meet the citation standards of the institution to which the work is being submitted. Failure to acknowledge an already existing work as your source, if it is your source, is plagiarism.

Plagiarism can take many forms and can occur in many contexts: a group of students working together, undergraduates using other people's work without acknowledging the sources, academic staff copying the work of graduate students, or researchers taking credit for work done by others in grant applications. Any instance of unacknowledged use of material produced by someone else is plagiarism.

4.2 Types of Plagiarism

There are countless instances of plagiarism in the world and classifying them all into neat categories is almost impossible. Sometimes, if a writer has inspired another writer, the inspiring writer can claim that the inspired writer has plagiarized. We will try to follow the types that have the broadest consensus among scholars.

Complete Plagiarism This type of plagiarism occurs when an author copies a complete paper or assignment and submits it as his or her own. It has some other names too: total or direct plagiarism. **Verbatim plagiarism** also falls under this category. If a complete work of another writer is copied as it exists and presented as one's own, it is verbatim plagiarism. This type of plagiarism is the most blatant violation of academic standards of ethical behavior and integrity and attracts the harshest forms of institutional penalties.

Mosaic or Patchwork Plagiarism This is the most problematic type of plagiarism because it mimics the same process which results in the production of systematic reviews, literature review, and surveys of existing research. There is only one characteristic that differentiates patchwork plagiarism from original research: the absence of attribution. If the writer adds quotation marks to indicate the borrowed sentences, provides the references to the sources, and offers additional commentary as a researcher, it will become a well-researched piece of writing. Patchwork plagiarism is as laborious as original research but it is categorized as plagiarism because it does not acknowledge its sources, does not display the ability to synthesize, and does not comment on the arguments presented.

According to Rebecca Moore Howard, "patchworking" is how researchers begin their training in any academic discipline and it can be a good way to begin to learn the craft of research and should not be treated as academic misconduct (Howard 1999). This perspective challenges traditional views of plagiarism and highlights the complexities involved in defining and addressing the issue of plagiarism. The writers of this book differ from the view of Rebecca Moore Howard (1999) because the intention of the plagiarist is not to appear as an apprentice researcher but as an accomplished researcher.

Paraphrasing Plagiarism occurs when a writer paraphrases the words of another writer without acknowledging the source(s). This kind of plagiarism only works with prose. It does not work with poetry because a poem has an underlying meter, the source of the cadence or rhythm in a verse. Changing a word in a poetic line changes the rhythm of the verse. Paraphrasing is an essential skill in all research writing. It becomes plagiarism only if the sources are not acknowledged.

Self-Plagiarism This type of plagiarism is considered a dubious practice because the writer tries to take the credit for one piece of writing multiple times. The legal principle behind this understanding of plagiarism stems from the fact that an act of productivity cannot be rewarded twice or thrice.

Universities have laid out rules to prevent this type of plagiarism: The same assignment cannot be submitted twice in two different courses, and the same dissertation cannot be submitted to obtain two degrees at two different universities. This type of plagiarism is a problem only in the context of academia and scientific research. Poets often recite the same poems at different reading venues, and singers often sing the same songs all over the world and gain admiration from new audiences. Anthologies and readers often include previously published articles, but the titles of the original publications remain unchanged. Self-plagiarism becomes a violation if an old work is published under a new title in a deceptive manner and the researcher tries to claim credit for producing a new piece of research.

We can illustrate this point with an example: Bapsi Sidhwa published a novel titled *Ice Candy Man* in England in 1988, but when it was published in the United States in 1991, it was titled *Cracking India*. It was not considered an act of self-plagiarism because everybody in the publishing industry knew it was the same novel with a different title. There was no deception involved. If she had tried to claim it as two different novels, she would have been accused of self-plagiarism. Self-plagiarism is all about intention.

Compilation Plagiarism This type of plagiarism takes place if a piece of research is full of citations and quotes, which are appropriately cited but, in the end, nothing new has been contributed to the existing knowledge in any academic discipline. This type of plagiarism is different from a systematic literature review and an annotated bibliography because it does not even offer any assessment of the cited literature. Etymologically speaking, this is the meaning of the original Latin word *plagiarius*. The words of other writers have been kidnapped and are held hostage and nothing new is happening.

Ghostwritten Work This practice is forbidden in an academic setting. Outside academia, it is quite common. Politicians and celebrities often hire ghostwriters to have their life stories narrated. They contribute the material and the writer turns it into a bestselling memoir. Speechwriters for various presidential candidates can change the political landscape of their country and the speaker is usually someone else. Barack Obama's speechwriter Cody Keenan is a respected visiting professor at Northwestern University, and he is referred to as Obama's collaborator. Hiring others to write for you becomes a problem if you are not ultrarich, famous, and powerful. If you are President Musharraf, it is not a problem (Ved 2006). Everybody assumes you have hired a ghostwriter.

Computer-Aided Plagiarism This matter is hotly debated. Computers do not have an intrinsic understanding of language. Artificial text-generating AI is itself built upon plagiarism. Many writers have already sued different AI companies for using their writings to train their large language models without seeking the consent. The entire body of writings done by human beings is plagiarized by machines which learn to regurgitate the words based on statistical relationships and probability. On the other hand, AI-based text generators are acting as a great equalizing force. Non-native speakers of English can get their grammar and sentences fixed by AI tools. So far, AI has not been able to come up with any original hypothesis about the nature of reality. Nevertheless, it is considered unethical to copy a piece of AI-generated text and to claim it as one's own. The authors of this book do not recommend this at all. It can be detected by an AI detector. It is interesting to note that when AI is writing and AI is detecting, how can human beings differentiate between an original piece of writing and an unoriginal piece of writing. AI is trained on human writing, on large language models (LLMs). Human beings are paying the cost of running the machines who are learning from and imitating their writings. If AI ever becomes fully conscious, it can be accused of inter-consciousness plagiarism.

Conceptual Plagiarism This type of plagiarism occurs when one researcher appropriates the ideas of another person and pretends to be the originator of those ideas. Often, the entire intellectual framework is hijacked and presented without any credit given to the original thinker. Conceptual plagiarism is difficult to detect if it is not accompanied with verbatim plagiarism. There are no global directories of ideas or frameworks to consult and verify who the original thinker of an idea is. This type of plagiarism is often the result of cultural appropriation. For example, Columbus is credited with discovering America yet the indigenous peoples already living in America are not regarded as discoverers. This leads us to cross-cultural plagiarism.

Cross-Cultural Plagiarism This type of plagiarism is very common. The discoveries of the global South are often attributed to the thinkers and artists of the global South. Pablo Picasso learned a great from African art (Cohen 2017) and it changed his artistic style and aesthetics, but he did not give the credit to African art. Cross-cultural plagiarism was particularly prevalent during the colonial era. In numerous cases, cultural artifacts were physically moved to the global North and placed in museums, a practice closely aligned with the etymological meaning of the word plagiarism.

4.3 Intentional and Unintentional Plagiarism

Plagiarism may be **intentional** or **unintentional** based on the intent of the author.

Intentional plagiarism is an author deliberately presenting somebody else's work or ideas (or their own prior work or ideas)—in part or whole, verbatim or paraphrased—as their own (or original).

Unintentional plagiarism is an author inadvertently presenting somebody else's work or ideas (or their own prior work or ideas)—in part or whole, verbatim or paraphrased—as their own (or original). Also called **accidental plagiarism,** it involves an author failing to realize that the idea has already been published or submitted by another or failing to use proper acknowledgment, attribution, or citation.

Accidental plagiarism often results from careless note-taking and sloppy work habits, although in some cases the author genuinely does not realize or remember how they came across the idea. Finally, though it is a rare occurrence, the same idea may appear to multiple people who are not in contact with one another.

It should be obvious from the above that while intentional plagiarism is a brazen act of academic dishonesty, unintentional plagiarism may result from the lack of diligence, carelessness, or a failure to follow citation guidelines of the institution receiving the submission.

What is less obvious is that unintentional plagiarism can be as injurious as intentional plagiarism to a scholar's academic worth and professional standing. In other words, there is little distinction between intentional and unintentional plagiarism as far as the consequences are concerned.

4.4 Why and How to Avoid Plagiarism

In scholarly circles, plagiarism is a major violation of standard practices because it is detrimental to the spirit of inquiry. The objective of academic work is to promote genuine scholarship and intellectual development and to push the frontiers of knowledge production and dissemination through teaching and publications. Plagiarism sabotages this mission by producing spurious forms of knowledge and phoney scholars who are ill equipped to pursue original research in any field of inquiry.

From a moral perspective, plagiarism, especially when intentional, is outright theft. Credit for genuine scholarship and original artistry is an intellectual property right protected in most modern jurisdictions. Plagiarism seeks to transfer the benefits of one person's intellectual exertions to another. Intellectual property is quite unlike other property in that its title cannot be transferred. Stripping a person of the title to their intellectual property is therefore unacceptable under any circumstances, not even on the pretext of payment or consideration. This is why gift authorship—adding someone's name to a research publication when they have not worked on it—is also considered a dubious activity.

Students must avoid plagiarism if they ever hope to 1 day become well-rounded scholars they aspire to be. You may perhaps breeze through your degree program by plagiarizing other people's work, but how will you fare in terms of intellectual development and moral integrity? If you are a student, you will emerge a much-diminished human being whose degree will scarcely be worth the paper it is printed on.

Here is a set of basic guidelines that students willing to avoid plagiarism may use to good effect. Feel free to build on them or develop your own protocols to avoid plagiarism.

Know Plagiarism The first step toward avoiding plagiarism is to familiarize yourself thoroughly with what constitutes plagiarism. You cannot steer clear of something unless you know what its nature is. Beyond that, it is a matter of following well-established research and writing practices and maintaining a stringent work ethic.

Cite Rigorously Academic institutions, without exception, require adherence to a set of accepted citation guidelines provided by various professional organizations: the American Psychological Association (APA Style), the Modern Languages Association (MLA Style), and The University of Chicago Press (The Chicago Manual of Style). Master at least one of the citation styles preferred by your institution and follow it rigorously to format your citations and references.

Document Faithfully Take comprehensive notes and also record the sources while doing this, keep them organized, and use them to substantiate your arguments. Keep a scratch pad handy and never procrastinate unless you want to learn the hard way how memory can play tricks on the lazy scholar.

Quote Scrupulously Never fail to use quotation marks to indicate directly copied text and always cite the source fully and correctly. Researchers tend to verify one another's sources.

Use Plagiarism Detection Tools Never shy away from passing your work through the crucible of cutting-edge software tools used to identify potential plagiarism issues. Turnitin and iThenticate are your best friends in this regard.

Review and Proofread Before you submit your work, be sure to review and proofread it to ensure that all sources used are properly cited as per your citation guidelines.

4.5 Correct Citation Practices

There is one rule of thumb in this area: Your reader should be able to locate your sources using the information you provide. Different institutions and professional bodies have developed various citation styles, but they all share common components. **For books**, the name of the author, the year of publication, the title of the book, the city of publication, and the name of the publisher are required. **For journal articles**, the name of the author, the year of publication, the article title, the journal title, the volume and issue numbers, the page range, the article covers, and, if available in digital format (usually PDF), the digital object identifier (DOI) are necessary. Older journal articles are also being assigned DOIs as journal databases are updated. **For online journalistic publications**, the name of the author, the year (including the month and day), the title of the article, the name of the newspaper or magazine, and the URL are required.

Every citation style requires these key components but the order in which this information is presented differs from one organization's style to another's. The citation styles of the Modern Languages Association (MLA) and the American Psychological Association (APA) no longer require the location of the publishing house. This is because most sources are published and accessed online, rendering the physical location of the publishing house irrelevant.

4.6 How to Cite

There are two components of a citation: in-text citations and an entry in the list of cited sources, also known as a bibliography or references, which is provided at the end of a scholarly work. You must use quotation marks if you are using someone else's words exactly (verbatim quotation). If you are summarizing or paraphrasing an idea, you must provide citations in your writing (in-text citations). The American Psychological Association requires you to include the author's surname, the year, and the page number (if applicable) in parentheses—for example a fictitious author named John Doe will be cited as, (Doe, 2024, p. 10)—to indicate the source of your idea or quotation. A page number is required to verify an exact quotation, but if you are discussing the entire publication, a page number is not necessary. For an online article not in PDF format, a page number is also not needed.

4.7 Identifying Citable and Uncitable Material

Theoretically, everything is citable. However, a strong intellectual argument is built on stable, reputable, and often peer-reviewed research-based publications.

Some types of sources are generally discouraged because they can change over time. If you base your argument on text that is likely to change or disappear, your argument is perceived as weak. For example, anyone can edit a Wikipedia page; therefore, it is not a reliable source on which to build your argument. A Facebook post can disappear at any time. A Facebook page can be suspended. However, a reputable publisher (e.g., Springer, Routledge, Wiley, or *The New York Times*) is unlikely to change what they have published. Even if revisions or corrections are necessary, they typically publish a revised version with a history of changes or errata clearly noted.

Everything that is permanently available is citable, including recordings of personal interviews with respondents or answers to questionnaires sent via email. Keep the recordings and emails secure in case an editor or reviewer wants to verify your sources. The American Psychological Association (APA) allows you to cite a personal interview in the text but not in the list of references, as this source is not publicly accessible. The Modern Languages Association (MLA), on the other hand, requires you to provide the details of the personal interview in the works cited list (e.g., Doe, John. Personal interview with the author. 11 October 2023).

You need to master the citation style preferred by the university in which you are enrolled or the publishing house to which you are submitting your writing. Citations allow you to stand on the shoulders of giants.

References

Cohen JI. Picasso and primitive art. Paris, Kansas City and Montreal. Burlingt Mag. 2017;159(1376):944–5.

Cresswell J. Plagiarism. In: Oxford dictionary of word origins. 2nd ed. Oxford University Press; 2010. p. 329.

Google. Google Books Ngram viewer. 2024; Retrieved November, 2024, from https://books.google.com/ngrams

Howard RM. Standing in the shadow of giants: plagiarists, authors, collaborators. Ablex Publishing; 1999.

Mann AJ. 'A mongrel of early modern copyright': Scotland in European perspective. In: Bently RA, Kretschmer M, editors. Privilege and property: essays on the history of copyright. Open Book Publishers; 2010. p. 51–66.

Seo JM. Plagiarism and poetic identity in martial. Am J Philol. 2009;130(4):567–93.

Shaw P. Plagiary. Am Scholar. 1982;51(3):325–37.

Taylor JS. The problematic etymology of "plagiarism": why inclusion requires this term's replacement. Equity Educ Soc. 2024; https://doi.org/10.1177/27526461241259015.

Ved M. Pen that fires. The Times of India. 2006, September 22.; https://timesofindia.indiatimes.com/edit-page/pen-that-fires/articleshow/2015426.cms.

5 Using Text-Matching Software to Deter Plagiarism: The Do's and Don'ts

Muhammad Salman Bashir

Plagiarism, the act of using someone else's work or ideas without proper acknowledgment, is a significant issue across academic, professional, and creative fields. Although a straightforward concept, plagiarism is defined and handled distinctively in different institutions. It has different manifestations, including direct plagiarism (copying text verbatim without citation), self-plagiarism (reusing one's previous work without disclosure), mosaic plagiarism (rephrasing someone else's work while retaining their ideas), and accidental plagiarism (unintentional failure to cite sources correctly) (Crowder et al. 2013). Plagiarism has wide-ranging repercussions and can damage reputations, lead to legal penalties, and diminish the credibility of academic and professional institutions. Even the renowned William Shakespeare has faced allegations of plagiarism. Using anti-plagiarism software, LaFayette College professor June Schlueter and independent scholar Dennis McCarthy discovered notable similarities between sections of Shakespeare's plays *Macbeth*, *King Lear*, *Henry VI*, and *Richard III* and an obscure 1576 unpublished manuscript titled *A Brief Discourse of Rebellion* by little-known writer George North (Walchover 2012).

The proliferation of digital content in the modern era has eased the replication of information and the act of "literary theft," raising concerns about plagiarism (Eret and Gokmenoglu 2010). Considering the academic setting, the University of Pretoria revealed that in a survey of 150 undergraduates, 80% admitted to regularly plagiarizing assignments from online sources (Russouw 2005). Plagiarism is frequently overlooked as a form of misconduct and many students don't perceive copying from the internet sources as a form of cheating (Carnero et al. 2017). In light of the increasing concern surrounding plagiarism, the accessibility of online sources has made it crucial to develop robust systems to detect and deter plagiarism efficiently.

M. S. Bashir (✉)
Department of Scientific Writing & Biostatistics, Research Services and Clinical Trial Administration – Research Center, King Fahad Medical City, Riyadh Second Health Cluster, Riyadh, Kingdom of Saudi Arabia
e-mail: mbashir@kfmc.med.sa

A. K. Abu-Shaheen et al. (eds.), *Introduction to Research Ethics and Academic Integrity*, https://doi.org/10.1007/978-981-96-9678-9_5

The increasing pressure to publish research has caused a surge in plagiarism incidents, creating a need for effective plagiarism detection softwares. In this regard, text-matching softwares have emerged as a valuable tool to combat plagiarism. Tools such as Turnitin, iThenticate, and Grammarly are widely used to identify similarities between a piece of writing and existing works (Anil et al. 2023). These platforms compare text against vast databases, including academic journals, books, websites, and other submissions. By flagging potential matches, they help educators, researchers, and professionals ensure that content is original or properly cited. These tools utilize algorithms that analyze text patterns to identify identical or similar sequences, and their databases are consistently refreshed to incorporate new publications and Web material. Although the innovation of text-matching softwares plays an essential role in promoting academic and professional integrity, it is not without limitations. The software serves as a detection mechanism, but human judgment is necessary to assess the context and determine whether flagged content constitutes plagiarism or appropriate use of source material. Moreover, these tools cannot detect deeper forms of plagiarism, such as idea theft or cleverly paraphrased content. Privacy concerns, data ownership, and the potential bias in algorithmic decision-making are significant factors that demand careful consideration when implementing these systems (Sain and Negi 2024). As the dependence on technology to promote academic and research integrity expands, it is crucial to strike a balance between using these tools as preventative measures and fostering an environment of ethical writing and responsible citation practices. Continuous innovation and adaptation are thus necessary to keep up with evolving plagiarism techniques.

This chapter highlights the effective implementation of text-matching softwares to reduce instances of plagiarism, outlining the effective strategies and common pitfalls as the "Do's and Don'ts." Additionally, it explores the ethical considerations used in such technology, underscoring the need for human oversight and employing these tools as educational resources rather than as disciplinary measures. The understanding of the Do's and Don'ts of text-matching software enables educators and professionals to foster an academic culture that prioritizes originality, nurtures integrity, and facilitates writers to become better and more responsible (Prashar et al. 2024).

5.1 The Use of Quotation Marks and Automatic Paraphrasing Tools

In academic and professional writing, the use of "quotation marks" and paraphrasing the original content are predominant techniques for referencing the contributions of other writers. Referencing is a formal method to deter plagiarism by acknowledging the work of others in research writing (Kent 2007). All researchers and students need to adhere to the general ethical principles of referencing sources and recognizing the fundamental values (fairness, trust, respect) and responsibility of academic integrity. Thus, the effective use of these techniques, such as quoting, paraphrasing, and summarizing, requires adherence to certain guidelines to ensure the originality of the final work and to prevent unintentional plagiarism. The proper use of quotation marks for verbatim text and avoiding the overreliance on automatic paraphrasing tools are therefore crucial elements in this context and are elaborated as follows (Roig 2018).

5.1.1 Adequate Use of Quotation Marks

Quoting original sources verbatim is an essential skill in academic writing, allowing the writer to incorporate specific ideas or statements directly from the source without misrepresentation. The correct use of quotation marks is critical for ensuring that readers clearly understand which parts of the text come from the original source. Verbatim quoting is particularly useful when exact wording is essential to preserve the meaning or when citing a notable statement or idea.

To correctly incorporate quotes, writers must follow a set of established guidelines. First, the quoted material should always be enclosed in quotation marks to distinguish it from the writer's original content. Secondly, proper attribution must be provided, including the source's author, title, and publication details, accompanied by an accurate citation format. This helps maintain transparency and allows readers to trace the original source of the quoted material. It is equally important to integrate the quote seamlessly into the text, ensuring it supports the writer's argument rather than standing alone without context.

Despite its usefulness, the overuse of direct quotes can present several pitfalls. Overreliance on quotes can overshadow the writer's voice, reducing the originality of the work. According to the book published by (Horkoff 2015), inexperienced writers often rely too heavily on direct quotations in research papers rather than taking the time to paraphrase the original material. At best, this diminishes the impact of the quotes; at worst, it creates a paper that appears hastily assembled from various external sources. Therefore, writers must use quotations sparingly for stronger influence (Horkoff 2015). Additionally, improper citation or failure to use quotation marks can lead to accusations of plagiarism. When opting to quote directly from a source, the following guidelines must be focused on the following:

- A direct quote should be used only when the original author has expressed an idea in such a precise and impactful way that rewording it would alter its intended meaning or tone.
- Always verify that you have copied the quotation exactly as it appears in the original source.
- Be sure to quote enough of the original text to accurately convey the author's viewpoint and clarify their perspective.
- Introduce the quote with a signal phrase (e.g., "As noted by Mayfair (2005)") to smoothly integrate it into your writing while also providing proper citation.
- Do not use quotes in isolation. Blend them into your own sentence structure to maintain a natural flow.

- If you need to remove or change words within a quote, do so carefully to avoid misrepresenting the original meaning. Edits should only be made for brevity or grammatical clarity.
- Use three dots [...] to indicate where you have left out part of the quote and four dots [....] if the omission includes a full sentence ending with a period, demonstrating thoughtful selection of the most relevant parts.
- Square brackets [] should be used to insert explanatory information or to adjust verb tense or wording for clarity within the quote.
- If the quote contains an error in spelling or grammar, insert [sic] immediately after the error to show that it appears in the source material and is not your own mistake.
- Enclose direct quotes in double quotation marks (" "), and use single quotation marks ('') for any quotes that appear within the quoted material itself.

Thus, writers must ensure that quotes are used sparingly and selectively, only when necessary to support key points, and should always be followed by the writer's analysis or interpretation. By balancing the use of quotes with original thought and analysis, writers can avoid common missteps and strengthen the integrity of their work.

5.1.2 Overreliance on Automatic Paraphrasing Tools

In recent years, automatic paraphrasing tools such as QuillBot, Spinbot, and other AI-powered systems have become increasingly popular among writers looking to rephrase text (Eka Fajar Rahmani 2023). These tools promise to simplify the paraphrasing process by automatically generating alternative phrasing for any given passage. Paraphrasing tools have gained popularity, especially in academia, because they play a significant role in enhancing the writing skills of students. The reason for this growing prominence is the increased expectations for higher writing standards in both academic and professional arenas (Eka Fajar Rahmani 2023). These tools have facilitated students in a time when plagiarism and incorrect citations are considered as a severe breach of academic conduct, hindering students' academic progress (Dinneen 2021). In some cases, interventions lead to unintended consequences. Some students have acknowledged the shortcomings of text-matching software and sought ways to bypass its effectiveness, an outcome that Comas (2020) characterizes as "cobra effect" (Comas-Forgas et al. 2021). This concept describes the unintended consequences that occur when efforts to solve a problem end up exacerbating it. When text-matching tools are used for discipline rather than educational support, it pushes students to find ways to manipulate the system. Literature manifests that when text-matching tools are used for discipline rather than educational support, it pushes students to find ways to manipulate the system. The cobra effect has given rise to three prevalent methods used to bypass plagiarism detection: (1) outsourcing work through contract cheating, (2) using machine translation, and (3) relying on automated paraphrasing tools (Mphahlele and McKenna 2019).

Although paraphrasing tools can be useful, overreliance on them poses significant risks, particularly when it comes to maintaining originality and avoiding plagiarism. If a student were to rewrite a text using automatic paraphrasing softwares and present it as their "original" document, it would still be considered as an act of academic cheating. According to Dickerson (2007), this constitutes a type of "facilitated plagiarism," as it involves incorporating third-party content and materials without appropriate acknowledgment. Automatic paraphrasing tools thus continue to support plagiaristic practices (Dickerson 2007).

One of the primary disadvantages of using automatic paraphrasing tools is the potential for unintentional plagiarism. These tools operate by altering sentence structure or swapping out words with synonyms, often without fully grasping the meaning or context of the original text. As a result, the paraphrased content may retain much of the original wording or structure, leading to a failure to sufficiently transform the source material. This can result in paraphrasing that risks being too similar to the original, making the writer susceptible to plagiarism accusations, even if the intention was to avoid copying (Kouam, A. W. F. 2024). From an ethical perspective, it is important to recognize the difference between paraphrasing and summarizing. Paraphrasing involves rewording the original content while preserving its meaning, whereas summarizing condenses the material to highlight only the key points. Both techniques require a strong personal interpretation, where the writer fully engages with the material and communicates it in their own words. Automatic tools can oversimplify this process, leading to mechanical alterations without genuine understanding. The end result is often a surface-level transformation that fails to reflect the writer's grasp of the content (Clarin et al. 2023).

To use paraphrasing tools effectively and ethically, writers should view them as starting points rather than final solutions. While these tools can offer suggestions for rewording, it is essential that writers manually revise the output, ensuring that the text reflects their own interpretation and understanding. Additionally, proper citation of the original source remains necessary, even if the text has been paraphrased. This practice helps maintain academic integrity and ensures that credit is given where it is due (Rogerson and McCarthy 2017). Writers are encouraged to combine the benefits of these tools with careful manual revisions to produce content that is both original and properly attributed.

5.2 Employing Plagiarism Detection Software

Since the year 2000, a wide range of new tools and technologies have been developed to identify plagiarism in both student writing and professional documents. Over the years, plagiarism detection software has become an indispensable tool in modern education, contemporary publishing, and professional writing. These tools have become a matter of public interest being instrumental in higher education policy. Plagiarism detection software offer a systematic way to detect the similarities and catch text matching of the submitted writing and vast databases of previously published materials in order to maintain the integrity of original work (Gupta 2017).

However, the existing plagiarism detection tools cannot always reliably identify plagiarism. While current software can flag potential instances of plagiarism through text matching, it cannot identify definitively whether plagiarism has actually occurred (Anson and Kruse 2023). They are unable to differentiate well-referenced similarities versus plagiarized ones. These factors have led to doubts and instigated debate on the use of the terms "plagiarism detection software" or "plagiarism checkers." Many authors have proposed alternate names for these tools that better reflect their functionality. For example, Foltýnek et al. (2019) proposed using alternative phrases such as "text-matching software" or "software that supports plagiarism detection," whereas Wikipedia opts for the term "content similarity detection" (Foltýnek et al. 2019). Furthermore, Weber-Wulff et al. (2013) criticized plagiarism detection tools, describing them as both "a crutch and a problem" and emphasized that they should not be seen as a definitive solution to the problem of plagiarism (Weber-Wulff et al. 2013).

5.2.1 Functional Specifications of Plagiarism Detection Software

Plagiarism detection tools function by comparing a submitted piece of text against extensive databases, which include published works, academic papers, and Web content. These tools scan the text for similarities by using sophisticated algorithms that analyze the structure, sentence patterns, and sequences of words. The goal is to identify any potential overlap with existing content and highlight areas where the text may be similar to previously published material (Sain and Negi 2024).

Several popular tools are widely used in both academic and professional settings. Turnitin is a well-known platform in education, frequently used by schools and universities to check student assignments. Grammarly offers plagiarism detection along with writing enhancement features, providing immediate feedback on sentence structure, grammar, and citation. iThenticate is favored by academic publishers for screening manuscripts, and Copyscape is popular among Web content creators and publishers to check for plagiarism across online platforms. Each of these tools has its strengths, with Turnitin offering in-depth integration with academic databases, while Grammarly and iThenticate excel at delivering user-friendly interfaces and access to a broad range of sources. According to Anson and Kruse (2023), Turnitin is often regarded as the leading example of plagiarism detection software, largely due to its widespread adoption by educational institutions across more than 100 countries. Initially created by iParadigms, a company founded by University of California, Berkeley, researchers, the software was sold to private investors in 2014 (Anson and Kruse 2023). Approximately 90% of universities in Australia and about 95% of higher-education institutions in the United Kingdom make use of Turnitin in some capacity (Silvey et al. 2016). In contrast, its usage in the United States is significantly lower, where plagiarism detection tools remain a contentious issue, particularly among writing studies scholars.

Plagiarism software contains several features that interact to analyze text input. These include:

- A text input area for users to submit their content.
- A preprocessing function that often involves converting document formats and extracting relevant information.
- A reference database or corpus of texts, which may also include access to search engines like Google.
- An algorithm that compares the submitted text against the corpus or online sources.
- A dashboard or interface displaying the similarity score or originality percentage, along with the number of matched sources.
- A feature that highlights all sections of the text-matching sources in the corpus, providing citations and links to the original material.

These tools come with certain limitations. For instance, while Turnitin and iThenticate can access extensive academic databases, they may not cover every published source, particularly paywalled content or offline materials. Moreover, each plagiarism detection software provides a similarity score or match percentage; a metric that indicates how much of the text overlaps with existing content. A high similarity score does not always indicate plagiarism and may just reflect the use of proper citation or common phrases. Therefore, it is crucial for users to recognize that these tools flag issues for review, but human interpretation remains mandatory.

5.2.2 Plagiarism Detection Software: Benefits and Limitations

The proactive use of plagiarism detection software offers numerous benefits, particularly in encouraging students, researchers, and authors to check their work before submission. These tools can serve as a preemptive step, allowing writers to identify any areas of concern in their work and address them prior to final submission. For students, this practice promotes a culture of academic honesty, where plagiarism detection software becomes an educational tool rather than a punitive measure. A study carried out at the University of Western Australia found that using text-matching software helped students learn to correctly cite sources and enhance their paraphrasing abilities.

In addition to proactive use, plagiarism detection software often includes feedback mechanisms that can serve as learning tools. Many of these tools highlight problematic sections, allowing writers to review flagged content and improve their writing quality, citation accuracy, creativity of new ideas, and paraphrasing skills (Sain and Negi 2024). Another significant benefit is that plagiarism detection software helps reduce unintentional plagiarism. Writers may sometimes unintentionally reproduce content from sources they have read or cite sources incorrectly due to misunderstanding citation practices. Detection software can highlight such overlaps, helping authors avoid accidental errors and ensuring that their work adheres to proper academic or professional standards. This ensures that even honest mistakes do not tarnish the credibility of their work. A pilot study by Youmans (2011) found out that students who were clearly informed about the use of the software tended to plagiarize less than those who were not warned (Youmans 2011).

While text-matching software is highly useful, it is not foolproof and comes with certain limitations. One of the most common issues is the occurrence of false positives, where the software flags text that is not actually plagiarized. For example, commonly used phrases, technical terminology, or properly cited material can sometimes trigger alerts. In some instances, text-matching software is used with thresholds that consider minimal text overlap as free of plagiarism, which can result in false negatives. Consequently, errors may include false positives, where students are wrongly accused of plagiarism, and false negatives, where instances of plagiarism go undetected. False negatives often occur because plagiarized content is altered or concealed in ways that prevent detection by the software. This can confuse users, especially if they do not understand the software's function or misinterpret the similarity score. In such cases, careful manual review is necessary to discern whether the flagged text truly indicates plagiarism or is simply a case of common wording.

Conversely, plagiarism detection software also struggles to identify certain forms of plagiarism, such as paraphrased ideas, translated content, or content from sources not included in the tool's database. Writers who paraphrase material effectively may still be reproducing someone else's ideas without giving proper credit, and while the wording may be different, the theft of intellectual property remains. Similarly, works translated from other languages may escape detection since the software focuses on textual similarities rather than the underlying ideas. Goddiksen et al. (2024) uncovered some of the limitations of text-matching software by highlighting that plagiarism does not necessarily involve copied text; it can also involve the incorrect use of ideas, images, or other elements taken from another person's work (Paludan Goddiksen et al. 2024).

Finally, there is a need to balance the use of plagiarism detection software with human oversight (Rodrigues et al. 2024). While these tools can flag potential issues, they cannot fully understand context, nuance, or academic intent. As mentioned earlier, manual review by instructors, editors, or peers is crucial in interpreting the results accurately and ensuring that genuine instances of plagiarism are identified and addressed appropriately. Overreliance on software alone can lead to unfair judgments or missed cases of plagiarism that require deeper investigation.

5.3 Ethical Considerations in the Application of Plagiarism Detection Software

The implementation of plagiarism detection software in educational and professional spheres raises several ethical considerations that must be addressed in order to ensure the responsible and fair use of these tools. It is thus mandatory to maintain a balance between academic integrity and punitive measures, privacy concerns, and the ethical challenges posed by automatic paraphrasing and text generation tools.

5.3.1 Academic Integrity Versus Punitive Measures

The ethical use of plagiarism detection tools should prioritize fostering academic integrity rather than serving as a punitive measure against those who may inadvertently violate citation practices. These tools should act as deterrents, promoting a culture of honesty and responsibility in writing. Instead of viewing these tools solely as mechanisms for catching wrongdoers, educators and institutions should embrace their role in enhancing understanding of proper citation and paraphrasing techniques. Roe and Perkins (2022) emphasized that text-matching software are a threat to educational integrity, and it is high time that training of students and faculty must be prioritized to reduce the use of automatic paraphrasing tools in all modern institutions (Roe and Perkins 2022).

Institutions should adopt a learning-focused approach, encourage improvement, and provide constructive feedback to students and authors, particularly when dealing with minor or unintentional infraction. By framing plagiarism detection as a learning opportunity rather than a reason for punishment, educators can help students grasp the importance of original work and proper attribution. This approach emphasizes personal growth and the development of writing skills rather than simply penalizing mistakes. This helps foster a culture of academic integrity, as incorporating plagiarism detection software as a learning aid shifts the educator's role toward supporting the development of writing skills rather than solely acting as an enforcer against plagiarism.

Clear communication with users, whether they are students, employees, or researchers, on how plagiarism detection software operates, what constitutes acceptable similarity levels, and the rationale behind its use is essential. Users should be informed about how their work will be analyzed, what actions may result from flagged content, and how to interpret similarity scores. This clarity fosters trust and helps users navigate the complexities of academic writing with confidence. This leaves a "take-home message" that a collective effort and commitment from authors, reviewers, editors, and policymakers are required to address the problem of plagiarism.

5.3.2 Privacy Concerns and Data Security

The ethical dilemmas associated with uploading student work or sensitive documents to third-party servers cannot be overlooked. When using plagiarism detection software, individuals often share their original content with external entities, raising significant privacy concerns. There is an inherent risk that sensitive or unpublished materials may be stored, accessed, or used in ways that the authors did not intend or anticipate. Plagiarism detection software is proven to be controversial due to ethical concerns about students' rights since these systems create permanent records of student work that can be reused for future plagiarism checks. Brinkman (2013) elucidated that students are entitled to clear information about how plagiarism detection operates and to know that their work will be stored permanently as part of the process (Brinkman 2013).

Legal considerations come into play, as institutions must ensure compliance with data privacy policies and international regulations, such as the General Data Protection Regulation (GDPR) in Europe. These regulations set stringent guidelines on how personal data can be collected, processed, and stored, requiring institutions to be vigilant in safeguarding users' information. Failure to comply with these regulations can lead to legal repercussions and loss of trust among users.

Best practices for mitigating privacy concerns include minimizing data retention and ensuring that students understand the implications of submitting their work to plagiarism databases. Institutions should establish clear policies regarding data storage, retention periods, and the potential for data deletion after certain timeframes. Providing transparency about how data will be used, shared, and protected is essential in reassuring users that their work is handled ethically and securely.

5.3.3 Ethical Challenges of Automatic Paraphrasing and Text Generation Tools

The rise of AI-generated content has introduced new ethical implications in academic writing, particularly concerning the use of automatic paraphrasing and text generation tools. While these tools can assist writers in generating ideas and rephrasing content, they often blur the line between legitimate paraphrasing and machine-generated text that lacks true originality or critical thought. Generative artificial intelligence tools, commonly known as text generation technologies, pose a threat to the credibility of the education system when misused. They can enable students to earn grades or qualifications they haven't genuinely earned. The rapid advancement of tools like ChatGPT has caught much of the academic community off guard, even though computer scientists have been refining such technologies for years. Clarke and Lancaster (2006) described automated academic text generation as an evolution of contract cheating practices (Clarke and Lancaster 2006).

One of the significant ethical challenges is the potential reliance on software-generated solutions instead of encouraging authentic engagement with the material. Wang (2024) carried out a thematic analysis of students' written work, reflections, and interviews, revealing that students used ChatGPT to help generate ideas, structure content, and address both broader writing concerns (such as argumentation, organization, and coherence) and more specific aspects (like grammar, word choice, and sentence structure). This suggests that relying heavily on automated tools may hinder writers from cultivating their own voice, critical thinking, and analytical abilities (Wang 2024). This can lead to a superficial understanding of the subject matter, as critical thinking and personal interpretation are essential components of academic writing.

Moreover, the reliance on these tools raises questions about the authenticity of work submitted. If a piece of writing is predominantly machine generated, does it genuinely reflect the author's understanding and voice? Educators and institutions must encourage users to cultivate original writing skills and engage deeply with their subjects rather than leaning on software to fulfill their writing requirements. Literature also reflects the use of ChatGPT in education as a curse; there is a degree of apprehension among concerned users about a potential overdependence of students and scholars on the AI system (Hadi Mogavi et al. 2024). Consequently, the reliance on AI-generated text and automatic text generation encourages superficial learning habits and erode students' social and critical thinking skills.

To address these ethical challenges, institutions should provide guidance on how to use automatic paraphrasing tools responsibly. This includes promoting the importance of personal interpretation and critical analysis while recognizing the limitations of AI-generated content. By emphasizing the value of original thought and authentic expression, educators can help students navigate the complexities of academic writing in an increasingly digital landscape. Hence, by presenting AI as a supportive educational tool rather than a shortcut, educators can foster a learning environment that emphasizes and nurtures critical thinking in the digital age.

5.4 The Do's and Don'ts of Text-Matching Software

The effective use of text-matching software in academic and professional contexts relies on understanding its strengths and limitations. Table 5.1 summarizes the key do's and don'ts that can guide users in leveraging these tools responsibly and ethically.

5.5 Conclusion

The use of text-matching software in academic and professional spheres represents a pivotal advancement in the ongoing battle against plagiarism. However, as we navigate this landscape, it is essential to emphasize the balance between leveraging technological tools and maintaining ethical writing practices. Text-matching software should be viewed as a means to foster academic integrity rather than merely a punitive tool aimed at identifying and penalizing infractions. By encouraging users to engage with these tools as part of their writing and learning process, educators can promote a culture of responsibility and respect for original work. When used effectively, these tools serve not only to identify potential issues but also to educate users about proper citation practices and the importance of maintaining their unique voices in their writing. Looking to the future, we can anticipate significant trends driven by advancements in artificial intelligence and machine learning. These technologies promise to enhance the capabilities of plagiarism detection tools, making them more sophisticated and accurate. As software evolves, it may increasingly integrate contextual understanding and semantic analysis, allowing for a more

Table 5.1 Highlighting the do's and don'ts of using text-matching software

Category	Do's	Don'ts
Plagiarism detection	Utilize text-matching software to spot and highlight possible plagiarism for further examination.	Depend entirely on software to detect all forms of plagiarism, including subtle rewording.
Educational use	Leverage flagged content to teach students proper citation practices.	Penalize students immediately without reviewing flagged content or giving feedback.
Fairness and transparency	Apply the software uniformly across all student submissions to ensure fairness.	Use different criteria or thresholds for various students, leading to potential bias.
Customization	Modify software settings to reflect the institution's academic policies.	Rely on default settings without considering how they may affect outcomes.
Human review	Pair software results with human insight for a better understanding of flagged content.	Rely only on the software's findings without human evaluation, especially in ambiguous cases.
Data security	Ensure student submissions and data are securely managed and meet privacy standards.	Neglect privacy and data protection requirements when handling student work.
Feedback mechanism	Provide comprehensive feedback based on software results, detailing areas for improvement.	Only give a similarity score without explaining its meaning or offering guidance.
Consistent updates	Keep the software current with updated databases and improved detection methods.	Use outdated software that might overlook newer forms of plagiarism or sources.
False positives	Review flagged similarities for context, like common phrases or correctly cited material.	Assume all flagged content is plagiarism without checking for common knowledge or proper citations.
Teaching academic integrity	Promote the use of text-matching software as part of teaching academic integrity.	Treat the software purely as a punitive measure, ignoring its educational benefits.

nuanced evaluation of text. This evolution will not only improve detection rates but also support writers in developing a deeper understanding of originality and ethical writing practices. Therefore, to successfully embrace the potential of text-matching software, it is crucial to prioritize educational objectives and ethical considerations. By doing so, we can create an environment that values integrity, fosters growth, and supports the continuous development of writing skills in an ever-evolving digital landscape.

References

Anil A, Saravanan A, Singh S, Shamim MA, Tiwari K, Lal H, Seshatri S, Gomaz SB, Karat TP, Dwivedi P, Varthya SB, Kaur RJ, Satapathy P, Padhi BK, Gaidhane S, Patil M, Khatib MN, Barboza JJ, Sah R. Are paid tools worth the cost? A prospective cross-over study to find the right tool for plagiarism detection. Heliyon. 2023;9(9):e19194. https://doi.org/10.1016/J.HELIYON.2023.E19194.

Anson CM, Kruse O. Plagiarism detection and intertextuality software. In: Digital writing technologies in higher education: theory, research, and practice. Cham: Springer; 2023. p. 231–43. https://doi.org/10.1007/978-3-031-36033-6_15.

Brinkman B. An analysis of student privacy rights in the use of plagiarism detection systems. Sci Eng Ethics. 2013;19(3):1255–66. https://doi.org/10.1007/s11948-012-9370-y.

Carnero AM, Mayta-Tristan P, Konda KA, Mezones-Holguin E, Bernabe-Ortiz A, Alvarado GF, Canelo-Aybar C, Maguiña JL, Segura ER, Quispe AM, Smith ES, Bayer AM, Lescano AG. Plagiarism, cheating and research integrity: case studies from a masters program in Peru. Sci Eng Ethics. 2017;23(4):1183–97. https://doi.org/10.1007/s11948-016-9820-z.

Clarin AS, Serohijos SS, Sumaylo LL, Baluyos GR. The challenges in paraphrasing among English language students. EduLine J Educ Learn Innov. 2023;3(4):493–503. https://doi.org/10.35877/454RI.EDULINE2052.

Clarke R, Lancaster T. Eliminating the successor to plagiarism? Identifying the usage of contract cheating sites. 2006.; https://www.researchgate.net/publication/228367576_Eliminating_the_successor_to_plagiarism_Identifying_the_usage_of_contract_cheating_sites.

Comas-Forgas R, Sureda-Negre J, Morey-López M. Spanish contract cheating website marketing through search engine advertisements. Assess Eval High Educ. 2021;46(7):1035–47. https://doi.org/10.1080/02602938.2020.1841091.

Crowder S, Shawn M, Turvey BE. Ethical issues for students of criminal justice. In: Ethical justice: applied issues for criminal justice students and professionals. Academic; 2013. p. 21–62. https://doi.org/10.1016/B978-0-12-404597-2.00002-4.

Dickerson D. Facilitated plagiarism: The Saga of term-paper mills and the failure of legislation and litigation to control them. Villanova Law Rev. 2007;52(1)

Dinneen C. Students' use of digital translation and paraphrasing tools in written assignments on Direct entry English Programs. Engl Aust J. 2021;37(1):40–53.

Eka Fajar Rahmani. Undergraduate students' perceptions on Quillbot paraphrasing tool. Scripta Engl Dept J. 2023;10(2):182–90. https://doi.org/10.37729/scripta.v10i2.3674.

Eret E, Gokmenoglu T. Plagiarism in higher education: A case study with prospective academicians. Procedia Soc Behav Sci. 2010;2(2):3303–7. https://doi.org/10.1016/J.SBSPRO.2010.03.505.

Foltýnek T, Meuschke N, Gipp B. Academic plagiarism detection. ACM Comput Surv (CSUR). 2019;52(6) https://doi.org/10.1145/3345317.

Goddiksen MP, Johansen MW, Armond AC, Centa M, Clavien C, Gefenas E, Kovács N, Merit MT, Olsson IA, Poškutė M, Santos JB. The dark side of text-matching software: worries and counterproductive behaviour among European upper secondary school and bachelor students. International Journal for Educational Integrity. 2024;20(1):15.

Gupta S. Plagiarism detection software: an overview. 2017.; https://www.researchgate.net/publication/335319583_Plagiarism_Detection_Software_an_Overview.

Hadi Mogavi R, Deng C, Juho Kim J, Zhou P, Kwon YD, Hosny Saleh Metwally A, Tlili A, Bassanelli S, Bucchiarone A, Gujar S, Nacke LE, Hui P. ChatGPT in education: a blessing or a curse? A qualitative study exploring early adopters' utilization and perceptions. Comput Hum Behav Artif Hum. 2024;2(1):100027. https://doi.org/10.1016/J.CHBAH.2023.100027.

Horkoff T. Chapter 9: Citations and referencing—Writing for success—1st Canadian Edition. 2015.; https://opentextbc.ca/writingforsuccess/chapter/chapter-9-citations-and-referencing/

Kent T. The complete guide to referencing and avoiding plagiarism. ELucidate. 2007;4(5):10.29173/ELUCIDATE449.

Kouam, A. W. F. (2024). What is plagiarism in academic writing and how to avoid it: Guiding novice researchers. International Journal of Research Publications, 144(1), 283–286.

Mphahlele A, McKenna S. The use of turnitin in the higher education sector: decoding the myth. Assess Eval High Educ. 2019;44(7):1079–89. https://doi.org/10.1080/02602938.2019.1573971.

Paludan Goddiksen M, Willum Johansen M, Catharina Vieira Armond A, Centa M, Clavien C, Gefenas E, Kovács N, Tang Merit M, Anna Olsson IS, Poškutė M, Borlido Santos J, Santos

R, Strahovnik V, Varga O, Wall PJ, Sandøe P, Bøker Lund T. The dark side of text-matching software: worries and counterproductive behaviour among European upper secondary school and bachelor students. Int J Educ Integr. 2024;20(1):1–18. https://doi.org/10.1007/S40979-024-00162-7.

Prashar A, Gupta P, Dwivedi YK. Plagiarism awareness efforts, students' ethical judgment and behaviors: a longitudinal experiment study on ethical nuances of plagiarism in higher education. Stud High Educ. 2024;49(6):929–55. https://doi.org/10.1080/03075079.2023.2253835.

Rodrigues M, Silva R, Borges AP, Franco M, Oliveira C. Artificial intelligence: threat or asset to academic integrity? A bibliometric analysis. Kybernetes. 2024; https://doi.org/10.1108/K-09-2023-1666.

Roe J, Perkins M. What are automated paraphrasing tools and how do we address them? A review of a growing threat to academic integrity. Int J Educ Integr. 2022;18(1):1–10. https://doi.org/10.1007/S40979-022-00109-W/METRICS.

Rogerson AM, McCarthy G. Using Internet based paraphrasing tools: original work, patchwriting or facilitated plagiarism? Int J Educ Integr. 2017;13(1):1–15. https://doi.org/10.1007/S40979-016-0013-Y/TABLES/2.

Roig M. Avoiding plagiarism, self-plagiarism, and other questionable writing practices: a guide to ethical writing. In: Getting to good: research integrity in the biomedical sciences. Springer New York University Langone Medical Center, NY, USA; 2018. p. 289–360.

Russouw R. Net closes on university cheats. 2005.; https://www.iol.co.za/news/south-africa/net-closes-on-university-cheats-235023.

Sain S, Negi A. Trends & challenges in plagiarism & text-matching software for academic integrity. Himalayiya University, Arunachal Pradesh, India; 2024.

Silvey V, Snowball T, Do T. Bridge over troubled water: a literacy approach to using Turnitin. J Acad Lang Learn. 2016;10(1):A206–22.

Walchover N. Plagiarism software pins down new source for Shakespeare's plays. The Guardian. 2012; https://www.theguardian.com/books/2018/feb/09/shakespeare-plagiarism-software-george-north

Wang C. Exploring students' generative AI-assisted writing processes: perceptions and experiences from native and nonnative English speakers. Technol Knowl Learn. 2024;1–22 https://doi.org/10.1007/S10758-024-09744-3/FIGURES/2.

Weber-Wulff D, Möller C, Touras J, Zincke E, Berlin H, Berlin HU. Plagiarism detection software test 2013. Plagiat.Htw-Berlin.De; 2013. p. 1–18.

Youmans RJ. Does the adoption of plagiarism-detection software in higher education reduce plagiarism? Stud High Educ. 2011;36(7):749–61. https://doi.org/10.1080/03075079.2010.523457

Responsible Scholarship

6

Caner Börekci

6.1 Introduction

In academic research, responsible scholarship is a crucial concept encompassing core values and practices designed to uphold scholarly work's integrity, reliability, and ethical standards. These principles—from research integrity and accountability to transparency and openness—are fundamental in producing credible and impactful research (Edwards 2002; Peels et al. 2019). A scholar who adheres to these standards contributes to the advancement of knowledge and reinforces the trust and respect that are essential within the academic and the general community.

A responsible scholar embodies several key characteristics that contribute to the integrity and quality of their research. Central to this is a commitment to research integrity, which involves honesty and accountability throughout the research process (Zahari et al. 2022). This encompasses accurate data collection, transparent reporting of results, minimizing bias, and avoiding manipulation (Weinbaum et al. 2019). By adhering to established professional norms and ethical principles, scholars ensure that their research practices meet high standards and contribute positively to society and the scientific community (Zhaksylyk et al. 2023). Accountability is closely intertwined with ethical integrity, requiring scholars to take full responsibility for their research activities (Valkenburg et al. 2021). This involves being ready to respond to questions or critiques, thoroughly documenting research procedures, and adequately crediting contributions in line with established standards.

Moreover, transparency is crucial in fostering trust within the academic community and the public. Responsible scholars conduct their research openly, disclosing their methods, materials, and any alterations made during the research process (Haven et al. 2022). This transparency facilitates the evaluation of research findings

C. Börekci (✉)
Research Center for IT, Balıkesir, Türkiye
e-mail: caner.borekci@balikesir.edu.tr

A. K. Abu-Shaheen et al. (eds.), *Introduction to Research Ethics and Academic Integrity*, https://doi.org/10.1007/978-981-96-9678-9_6

and minimizes duplication, especially in fields like medical research, where precision and clarity are paramount (Bradley et al. 2020). In addition to practicing transparency, responsible scholars prioritize objectivity as a cornerstone of their work (Closa 2021). They actively strive to minimize bias in all aspects of research, from experimental design to data interpretation (MacCoun 1998; Resnik 2014). This effort involves being conscious of personal and financial interests that could influence research outcomes, committing to fair and impartial peer review processes. Responsible scholars demonstrate their dedication to objectivity by complementing it with carefulness (Zhaksylyk et al. 2023). They thoroughly examine their work and that of their peers, maintain detailed records of research activities, and rigorously evaluate all findings before publication (Kouam 2024). By doing so, they prevent negligence and ensure the reliability and accuracy of research outcomes. Finally, open-mindedness is essential for fostering innovation and collaboration within the research community. Scholars must be receptive to new ideas, constructive feedback, and differing viewpoints, all of which can enhance the research process and more robust outcomes (Snow 2022).

In addition to these qualities, a responsible scholar is dedicated to continuous learning and professional development in ethical research practices. They stay informed about institutional guidelines, federal regulations, and the ethical implications of their methodologies. By adhering to the best research ethics and integrity practices, they maintain the high standards necessary for responsible scholarship. Equally important is their dedication to promoting fairness and justice, which involves respecting the rights of research subjects, ensuring the humane treatment of all participants, and upholding principles such as beneficence and justice in research design and implementation.

From this point of view, as a responsible scholar, developing good research practices and conducting research involves several key considerations as outlined in Fig. 6.1.

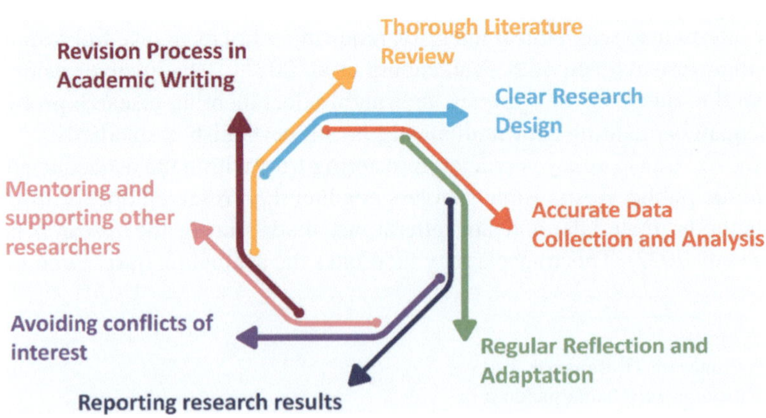

Fig. 6.1 Six key considerations to maintain a responsible scholarship

Thorough Literature Review To achieve a comprehensive understanding of the current state of knowledge in the field, a thorough literature review is conducted and involves several critical steps (Denney & Tewksbury 2012; Snyder 2019). The process begins with clearly defining the scope and objectives of the research (Rowley & Slack 2004), including the key themes, time frame, and geographical focus. It is crucial to search for relevant literature through academic databases, journals, books, and credible online sources, utilizing appropriate keywords and filters to ensure exhaustive coverage of the topic (Denney & Tewksbury 2012). After gathering the relevant literature, organize and categorize it according to themes, methodologies, or chronological order (Templier & Paré 2015). Using citation management tools like EndNote, Mendeley, or Zotero can facilitate this organization (Mahajan & Hogarth 2013). The next step involves critically analyzing each source to assess its validity, reliability, and relevance to the research. This analysis aids in identifying key arguments, methodologies, findings, and limitations within the existing body of work (Paré et al. 2015). Synthesizing this information allows researchers to identify knowledge gaps, emerging trends, and unresolved questions, which can then inform the development of a theoretical framework or conceptual model to guide the research. When writing the literature review, it is essential to structure the discussion thematically, methodologically, or chronologically, depending on the research focus (Synder 2019). The review should clearly articulate how the study fits into the existing body of knowledge, addresses identified gaps, and contributes to the field while providing a balanced analysis that acknowledges the literature's strengths and limitations (Rowley & Slack 2004; Snyder 2019). Upon completion, the literature review should be carefully revised to ensure clarity, coherence, and academic rigor. Feedback from peers, mentors, or advisors should be sought as necessary (Knopf 2006). Throughout the process, it is crucial to ensure that all sources are correctly cited and to maintain a complete and well-organized bibliography. Finally, the literature review should be periodically updated to include new findings, ensuring the research remains up to date. In addition, conducting a literature review is essential for maintaining academic integrity. Scholars must ensure accurate citations, avoid plagiarism, and represent research findings truthfully (Heston 2023). Respect for intellectual property, transparency, and openness are also crucial. Consider bias and objectivity when selecting sources, and handle sensitive data carefully. By adhering to these principles, researchers can contribute to a rigorous and trustworthy literature review that advances knowledge in their field.

Clear Research Design A clear and robust research design should be developed, incorporating well-defined objectives, methodologies, and analysis strategies (Creswell & Creswell 2017). It must be tailored to answer the specific research questions relevant to the field of interest. Additionally, the research design must rigorously adhere to ethical guidelines. Ensuring the protection of participants, maintaining transparency in data collection and analysis, and upholding the integrity of the research process are essential aspects (Aguinis & Henle 2004). Ethical considerations must be integrated from the formulation of research questions to the reporting of results, to ensure the trustworthiness and credibility of the study. These considerations include obtaining informed consent from participants, ensuring confidentiality when necessary, and avoiding any form of manipulation that could compromise the validity of the findings (Aguinis & Henle 2004).

Accurate Data Collection and Analysis Accurate data collection and analysis are imperative for rigorous research. Implementing meticulous data collection methods, maintaining comprehensive records, and utilizing validated tools are essential practices that enhance the reliability and validity of research findings (Hardwicke et al. 2018; Mark et al. 1999). Ensuring transparency in the research process further strengthens its credibility. The accuracy of data collection and analysis goes beyond mere methodological considerations; it constitutes a fundamental ethical responsibility. Ethical research requires that data be collected honestly and without bias, with full respect for the rights and dignity of all participants (Cohen et al. 2013). This includes obtaining informed consent, safeguarding participant confidentiality, and strictly avoiding any form of data fabrication or manipulation.

Maintaining detailed records is essential for both the reproducibility of research and the independent verification of findings. Comprehensive documentation facilitates a transparent research process, enabling peers to review and scrutinize each step, upholding the ethical principles of transparency and accountability (Kapiszewski & Karcher 2021; Tuval-Mashiach 2017). The use of validated tools and methods is also a key ethical requirement, safeguarding the accuracy of data collection and ensuring that the analysis reflects the true nature of the phenomena being studied. This approach mitigates the risk of misleading conclusions, which could have significant ethical implications, mainly when the research influences policy, practice, or public opinion. Process transparency is equally crucial from an ethical perspective, allowing others to comprehend and trust research findings. Openly sharing study methods, analysis procedures, and limitations fosters a culture of honesty and integrity in academic research (Kapiszewski & Karcher 2021). By committing to accurate data collection and analysis within an ethical framework, researchers contribute to developing reliable, valid, ethically sound, and socially responsible knowledge.

Regular Reflection and Adaptation Regular reflection and adaptation are essential to a rigorous research process. Ongoing evaluation and a willingness to refine methods through seeking feedback from peers and making necessary adjustments serve to enhance the quality and rigor of the research (Brandon et al. 2015). Regular reassessment of methods allows for identifying and resolving ethical concerns, such as potential biases, conflicts of interest, or unintentional harm to participants. Seeking peer feedback is an ethical practice that invites constructive scrutiny to ensure the rigor and fairness of the research. Openness to constructive criticism and a readiness to adjust methods demonstrate a commitment to the ethical principles of honesty, integrity, and transparency. Adapting research methods in response to ethical considerations might involve revising data collection techniques to improve participant confidentiality, adjusting analysis strategies to avoid misrepresenting findings, or even reconsidering the scope of the research if the original design poses ethical dilemmas. By integrating ethical reflection into the research process, researchers uphold their responsibility to conduct scientifically sound and ethically responsible studies (National Academies of Sciences 2017).

Reporting Research Results Ethically reporting results is fundamental to research ethics, ensuring scientific work's integrity, transparency, and credibility (Kambhampati et al. 2023). An accurate and honest representation of findings is essential for maintaining research integrity and preventing misconduct, such as data falsification, fabrication, or selective reporting (Aguinis & Henle 2004). This accuracy not only preserves the credibility of the research but also upholds the researcher's reputation. Transparency in reporting fosters open communication by fully disclosing all relevant data, methods, and results. This practice enables other researchers to critically evaluate and replicate studies, thereby contributing to the advancement of knowledge. Ethical reporting also entails disclosing limitations and conflicts of interest, which provides essential context for assessing the validity of research. Moreover, it fosters public trust and accountability within the scientific community. It ensures that research findings are reliable and based on accurate data. This accountability promotes a culture of mutual respect and shared ethical standards in academia (Moher et al. 2020). Finally, ethical reporting safeguards research participants and stakeholders by ensuring their contributions are respected and used responsibly (Cohen et al. 2013). It helps provide accurate information for evidence-based decision-making. In summary, ethically reporting research results is essential for upholding scientific work's integrity, transparency, and credibility.

In addition to these considerations, scholars may collaborate with colleagues or organizations throughout the research process. This collaboration necessitates that scholars assume a range of responsibilities. Some of these responsibilities include avoiding conflicts of interest, mentoring and supporting fellow researchers, and having their research reviewed and revised by peers.

Avoiding Conflicts of Interest A conflict of interest in academic research arises when a researcher's personal, financial, or professional interests can influence, or appear to influence, the research process or its outcomes. It undermines objectivity, integrity, and credibility, potentially leading to biased results or unethical practices. Personal interests, like relationships or beliefs, may cause unconscious and selective bias, leading researchers to favor specific outcomes (Rohwer et al. 2017). Financial interests, such as funding from entities with vested interests, can pressure researchers to align results with funders' goals. Researchers with financial stakes in companies or products related to their work may prioritize personal gain over scientific integrity (Dunn et al. 2016). Similarly, professional interests, like career advancement, may drive unethical practices, including selectively reporting data to achieve favorable results. Institutional interests further complicate the issue. Academic institutions with commercial partnerships or funding agreements may pressure researchers to produce results that align with the institution's goals, compromising research independence (Resnik 2015). The consequences of conflicts of interest are significant. They can result in biased research that favors the researcher's or affiliates' interests, undermining credibility of both the findings and the researchers. In extreme cases, they may lead to ethical violations like data manipulation or plagiarism. To maintain academic integrity, researchers must disclose potential conflicts of interest to their institutions and funding agencies. This transparency fosters critical evaluation and helps preserve trust. Many institutions also implement policies and oversight to manage these conflicts, ensuring research remains ethical and unbiased (Resnik et al. 2016).

Mentoring and Supporting Other Researchers Mentoring is vital for promoting academic integrity by fostering ethical conduct, adherence to research standards, and responsible scholarship. Mentors serve as role models, demonstrating ethical behavior throughout the research process, from forming questions to publishing results (Ochroch & Eckenhoff 2011). Early-career researchers learn by observing and internalizing these practices. Mentors also help navigate ethical dilemmas, reinforcing integrity in decision-making. Mentorship emphasizes best practices, including rigorous data collection, accurate analysis, and transparency. Mentors enhance the credibility of individual studies and the broader academic community by promoting the disclosure of methods, data, and potential conflicts of interest (Ochroch & Eckenhoff 2011). Mentoring plays a crucial role in instilling a sense of responsibility in emerging researchers, guiding them to uphold ethical standards and fostering a culture of accountability, both for themselves and their peers.

Furthermore, mentoring helps prevent research misconduct. Close supervision allows mentors to detect issues like data fabrication or plagiarism early, preventing escalation. This support system gives less experienced researchers confidence to ask questions, reducing the likelihood of unethical practices. Mentoring ensures the continuity of ethical scholarship across generations, preserving the long-term integrity of academic research. By guiding the next generation, mentors contribute to a supportive academic community where ethical behavior is encouraged and valued (Eby et al. 2013). To conclude, mentoring is essential for enhancing academic integrity by transmitting ethical standards, promoting best practices, preventing misconduct, and ensuring the sustainability of responsible research.

Revision Process in Academic Writing The revision process in academic writing is crucial for upholding research ethics by ensuring the work's accuracy, clarity, and integrity. Revisiting a manuscript allows researchers to verify data and validate findings, reducing errors and ensuring reliability (Misra et al. 2018). This process also helps prevent plagiarism by ensuring proper citation, accurate paraphrasing, and intellectual honesty. Revision reinforces transparency and honesty, ensuring the methods, results, and conclusions are clearly and accurately presented (Jana 2019). It also helps authors clearly outline the limitations of their research, fostering trust with readers. Additionally, during revision, researchers can assess whether their interpretations were fair and balanced, avoiding exaggeration or minimizing the significance of their findings. This step includes disclosing potential conflicts of interest to maintain objectivity and credibility (Makarem et al. 2023).

Respect for research participants is also reinforced through revision by ensuring confidentiality whenever necessary, particularly in studies involving sensitive data. This safeguards participants' rights and privacy, ensuring compliance with ethical guidelines (Cohen et al. 2013). The revision process further ensures adherence to ethical guidelines established by academic institutions, journals, and professional organizations. Researchers can confirm that all contributors are properly credited and that authorship criteria are met, promoting ethical authorship practices. In conclusion, revision is vital to research ethics, enhancing accuracy, transparency, honesty, and ethical reporting. Through thorough revision, researchers improve the quality of their work while upholding the ethical standards essential to academic research.

6.2 Conclusion

Responsible scholarship is the cornerstone of academic research, requiring adherence to principles of integrity, transparency, accountability, and ethical conduct. By committing to these values, scholars contribute to the advancement of knowledge while fostering trust and respect within the academic community. A thorough literature review, a clear research design, accurate data collection, regular reflection, and the ethical reporting of results are essential to maintaining scholarly responsibility. Furthermore, scholars must declare conflicts of interest, mentor emerging researchers, and engage in a rigorous revision process to ensure the credibility and ethical rigor of their work. Through these practices, responsible scholarship upholds the standards necessary for credible, reliable, and impactful research that benefits both academia and society at large.

References

Aguinis H, Henle CA. Ethics in research. In: Rogelberg SG, editor. Handbook of research methods in industrial and organizational psychology. Blackwell Publishing; 2004. p. 34–56. https://doi.org/10.1002/9780470756669.ch2.

Bradley SH, DeVito NJ, Lloyd KE, Richards GC, Rombey T, Wayant C, Gill PJ. Reducing bias and improving transparency in medical research: a critical overview of the problems, progress and suggested next steps. J R Soc Med. 2020;113(11):433–43. https://doi.org/10.1177/0141076820956799.

Brandon C, Jamadar D, Girish G, Dong Q, Morag Y, Mullan P. Peer support of a faculty "writers' circle" increases confidence and productivity in generating scholarship. Acad Radiol. 2015;22(4):534–8. https://doi.org/10.1016/j.acra.2014.12.006.

Closa C. Planning, implementing and reporting: increasing transparency, replicability and credibility in qualitative political science research. Eur Political Sci. 2021;20(2):270–80. https://doi.org/10.1057/s41304-020-00299-2.

Cohen L, Manion L, Morrison K. The ethics of educational and social research. In: Research methods in education. Routledge; 2013. p. 99–128. https://doi.org/10.4324/9780203720967.

Creswell JW, Creswell JD. Research design: qualitative, quantitative, and mixed methods approaches. Sage; 2017.

Denney AS, Tewksbury R. How to write a literature review. J Crim Justice Educ. 2012;24(2):218–34. https://doi.org/10.1080/10511253.2012.730617.

Dunn AG, Coiera E, Mandl KD, Bourgeois FT. Conflict of interest disclosure in biomedical research: a review of current practices, biases, and the role of public registries in improving transparency. Res Integr Peer Rev. 2016;1(1):1–8. https://doi.org/10.1186/s41073-016-0006-7.

Eby LT, Allen TD, Hoffman BJ, Baranik LE, Sauer J, Baldwin S, et al. An interdisciplinary meta-analysis of the potential antecedents, correlates, and consequences of protégé perceptions of mentoring. Psychol Bull. 2013;139(2):441–76. https://doi.org/10.1037/a0029279.

Edwards A. Responsible research: ways of being a researcher. Br Educ Res J. 2002;28:157–68. https://doi.org/10.1080/01411920120122121.

Hardwicke T, Mathur M, MacDonald K, Nilsonne G, Banks G, Kidwell M, et al. Data availability, reusability, and analytic reproducibility: evaluating the impact of a mandatory open data policy at the journal cognition. R Soc Open Sci. 2018;5(8):180448. https://doi.org/10.1098/rsos.180448.

Haven T, Gopalakrishna G, Tijdink J, van der Schot D, Bouter L. Promoting trust in research and researchers: how open science and research integrity are intertwined. BMC Res Notes. 2022;15(1):302. https://doi.org/10.1186/s13104-022-06169-y.

Heston TF. Foundations of scholarly writing. BP International; 2023. https://doi.org/10.2139/ssrn.4653380.

Jana S. A history and development of peer-review process. Ann Libr Inf Stud. 2019;66(4):152–62.

Kambhampati SBS, Menon J, Maini L. Ethics in research and publications. JOIO. 2023;57:1722–34. https://doi.org/10.1007/s43465-023-00971-x.

Kapiszewski D, Karcher S. Transparency in practice in qualitative research. PS Polit Sci Polit. 2021;54(2):285–91. https://doi.org/10.1017/S1049096520000955.

Knopf JW. Doing a literature review. PS Polit Sci Polit. 2006;39(1):127–32. https://doi.org/10.1017/S1049096506060264.

Kouam A, W., F. Mastering validity and reliability in academic research: meaning and significance. Int J Res Publ. 2024;144(1):287–92. https://doi.org/10.47119/ijrp1001441320246160.

MacCoun RJ. Biases in the interpretation and use of research results. Annu Rev Psychol. 1998;49(1):259–87. https://doi.org/10.1146/annurev.psych.49.1.259.

Mahajan AK, Hogarth DK. Taking control of your digital library: how modern citation managers do more than just referencing. Chest. 2013;144(6):1930–3. https://doi.org/10.1378/chest.13-0856.

Makarem A, Mroué R, Makarem H, Diab L, Hassan B, Khabsa J, Akl EA. Conflict of interest in the peer review process: a survey of peer review reports. PLoS One. 2023;18(6):e0286908. https://doi.org/10.1371/journal.pone.0286908.

Mark MM, Eyssell KM, Campbell B. The ethics of data collection and analysis. N Dir Eval. 1999;82:47–56. https://doi.org/10.1002/ev.1136.

Misra DP, Ravindran V, Agarwal V. Integrity of authorship and peer review practices: challenges and opportunities for improvement. J Korean Med Sci. 2018;33(46):e287. https://doi.org/10.3346/jkms.2018.33.e287.

Moher D, Bouter L, Kleinert S, Glasziou P, Sham MH, Barbour V, et al. The Hong Kong principles for assessing researchers: fostering research integrity. PLoS Biol. 2020;18(7):e3000737. https://doi.org/10.1371/journal.pbio.3000737.

National Academies of Sciences, Policy, Global Affairs, & Committee on Responsible Science. Fostering integrity in research. National Academies Press; 2017. p. 10.17226/21896.

Ochroch EA, Eckenhoff RG. The role of mentoring in aiding academic integrity. Anesth Analg. 2011;112(3):732–4. https://doi.org/10.1213/ane.0b013e318209d4e8.

Paré G, Trudel MC, Jaana M, Kitsiou S. Synthesizing information systems knowledge: a typology of literature reviews. Inf Manag. 2015;52(2):183–99. https://doi.org/10.1016/j.im.2014.08.008.

Peels R, de Ridder J, Haven T, Bouter L. Value pluralism in research integrity. Res Integr Peer Rev. 2019;4:1–13. https://doi.org/10.1186/s41073-019-0076-4.

Resnik DB. Bias, misconduct, and integrity in scientific research. In: The Routledge companion to bioethics. Routledge; 2014. https://doi.org/10.4324/9780203804971.

Resnik DB. Institutional conflicts of interest in academic research. Sci Eng Ethics. 2015;25(6):1661–9. https://doi.org/10.1007/s11948-015-9702-9.

Resnik DB, Ariansen JL, Jamal J, Kissling GE. Institutional conflict of interest policies at US academic research institutions. Acad Med. 2016;91(2):242–6. https://doi.org/10.1097/ACM.0000000000000980.

Rohwer A, Young T, Wager E, Garner P. Authorship, plagiarism and conflict of interest: views and practices from low/middle-income country health researchers. BMJ Open. 2017;7(11):e018467. https://doi.org/10.1136/bmjopen-2017-018467.

Rowley J, Slack F. Conducting a literature review. Manag Res News. 2004;27(6):31–9. https://doi.org/10.1108/01409170410784185.

Snow NE. The value of open-mindedness and intellectual humility for interdisciplinary research, Sci Fides. 2022;10(2):51–67. https://doi.org/10.12775/setf.2022.018.

Snyder H. Literature review as a research methodology: an overview and guidelines. J Bus Res. 2019;104:333–9. https://doi.org/10.1016/j.jbusres.2019.07.039.

Templier M, Paré G. A framework for guiding and evaluating literature reviews. Commun Assoc Inf Syst. 2015;37:112–37. https://doi.org/10.17705/1CAIS.03706.

Tuval-Mashiach R. Raising the curtain: the importance of transparency in qualitative research. Qual Psychol. 2017;4(2):126–38. https://doi.org/10.1037/qup0000062.

Valkenburg G, Dix G, Tijdink J, de Rijcke S. Expanding research integrity: a cultural-practice perspective. Sci Eng Ethics. 2021;27(1):10. https://doi.org/10.1007/s11948-021-00291-z.

Weinbaum C, Landree E, Blumenthal MS, Piquado T, Gutierrez CI. Ethics in scientific research. RAND Corporation; 2019.

Zahari AI, Said J, Arshad R. Examining the components of integrity. Integr Psychol Behav Sci. 2022;234-265 https://doi.org/10.1007/s12124-021-09626-8.

Zhaksylyk A, Zimba O, Yessirkepov M, Kocyigit BF. Research integrity: where we are and where we are heading. J Korean Med Sci. 2023;38(47):e405. https://doi.org/10.3346/jkms.2023.38.e405.

Predatory Dissemination of Research Findings

7

Amani Khalil Abu-Shaheen

7.1 Predatory Dissemination of Research Findings

The interdependence between scholarly and commercial publishing makes the publishing industry vulnerable to infiltration by unscrupulous publishers. Researchers depend heavily on academic publishers to disseminate their research findings. Equally, publishers rely on researchers as their main source of income. In 2021, the global scholarly publishing market was estimated at approximately $28 billion, with scientific journal publishing accounting for around $10–12 billion annually, dominated by a few major commercial publishers (Research and Markets 2021; Ware and Mabe 2021). Consequently, unscrupulous publishers target unsuspecting clients who wish to disseminate their research findings only to offer substandard services (Cobey et al. 2018; Grudniewicz et al. 2019).

7.2 Origins of Predatory Publishing and Beall's List

Predatory publishers have existed for some time since the onset of online publications. One of the first individuals to take note of the predatory publication patterns was Jeffrey Beall. In 2009, Jeffrey Beall, a scholarly librarian, began compiling and publicly sharing a list of publishers he described as "predatory open-access publishers," ultimately known as Beall's list. These entities exploited the author-pays model of open-access publishing by prioritizing profit over scholarly integrity. They often failed to provide legitimate editorial and peer-review services, despite claiming to do so. Beall's primary concern was that such publishers undermined the credibility of open-access publishing and posed a threat to academic integrity (Beall 2012,

A. K. Abu-Shaheen (✉)
Research Center, Riyadh Second Health Cluster, Riyadh, Kingdom of Saudi Arabia

A. K. Abu-Shaheen et al. (eds.), *Introduction to Research Ethics and Academic Integrity*, https://doi.org/10.1007/978-981-96-9678-9_7

2015). Until 2017, Beall kept updating his list of suspected predatory publishers to facilitate cross-referencing for researchers.

In his updates, Beall noticed that predatory journals grew exponentially at alarming rates of 187% and 82% (Beall 2015; Eriksson and Helgesson 2017). Moreover, research published in Science in 2013 confirmed Beall's suspicions (Bohannon 2013). In the article, the acceptance rate of submitted journals by predatory publishers exceeded the average acceptance rate of legitimate publishers. Legitimate publishers accepted 38% of submissions after thorough peer review, compared to the 68% acceptance cut-off by predatory publishers, confirming the substandard services offered by such publishers. To minimize susceptibility to predatory publishers, Beall created criteria alongside the list to characterize the activities of predatory publishers.

Other scientists and researchers have also conducted several experiments to expose predatory publishers. In 2008, Gunther Eysenbach drew attention to fraudulent publishers, particularly within the open-access domain (Eysenbach 2008). That same year, Richard Poynder highlighted the risks posed by unscrupulous publishers exploiting open-access journals for profit (Poynder 2008). In 2010, Phil Davis, a graduate student from Cornell University, created a manuscript using the computer program SCigen. He then submitted the manuscript, which had random, meaningless text, to one of these dubious publishers, and, predictably, the site accepted it for a small fee. However, Davis ultimately withdrew the submission, exposing the publisher's lack of scrutiny (Davis 2010).

In 2013, John Bohannon created a flawed document to test the authenticity of various publishers, known as the Bohannon experiment (Bohannon 2013). As expected, 60% of the potential predatory journals accepted the script, with 40% of the publishers rejecting the fake medical paper (Bohannon 2013). In 2015, four scientists created a counterfeit and substandard scientist whom they dubbed Dr. Anna Szust. The scientists then made applications for the editorial posts in 360 scholarly journals (Grudniewicz et al. 2019). A third of the 360 journals were from Beall's list of predatory journals, while the rest were legitimate journals. The scientists created false book chapters under her name and fake publishing houses through which she published. Although Dr. Szust had zero experience as an editor, 40 of the predatory journals accepted her offer without doing the bare minimum vetting process for the candidate. By contrast, the authentic journals offered minimal to no attention to Dr. Szust's application, implying they carried out a rigorous vetting process (Grudniewicz et al. 2019).

In another experiment, three students from the Massachusetts Institute of Technology (MIT) used information technology to expose predatory publishers. The three students worked in the computer science and artificial intelligence lab (CSIL), where they generated some nonsensical journals using the SCigen tool. The students created a paper accepted into the World Multi-Conference on Systemic Cybernetics, which was already suspected as predatory (Van Noorden 2014).

7.3 Beall's Criteria and Characteristics of Predatory Publishers

Drawing from his experience as a librarian, Beall created a standard criterion of 56 features that could be applied to scrutinize open-access publishers to expose predatory patterns (Beall 2015). Moreover, Beall drew insights from the Committee on Publication Ethics affiliated with the Open Access Scholars Publishers Association (Committee on Publication Ethics [COPE] 2011). One of the features used to identify predatory publishers is the absence of editors. Most predatory journals either have no list of editors or have a single board with unverifiable editors. Beall also noted that most predatory journals consisted of members with no academic expertise in published journals (Abalkina 2021a; Editage Insights 2021). A perfect example is a recent boom in periodicals during the COVID-19 era. Several predatory journals have a background in linguistics. Anna Abalkina, a scholar at the Free University of Berlin, noticed a dubious trend where three predatory publishers appeared in several research articles and journals on COVID-19, nutrition, and other medical fields. However, they had no affiliation with the medical field. The findings by Abalkina solidified the second criterion as listed by Beall (Abalkina 2021b).

Another distinguishing characteristic of the predatory journal is the lack of clear information concerning editing fees. Predatory publishers often lack a transparent fee structure and fail to properly index submitted papers, undermining the integrity of the publishing process. Another notable characteristic of predatory journals is their misleading titles. Often, the names of these journals bear little relation to their stated mission or editorial focus. For instance, a journal may bear the name American, while none of its editors or publishers have any academic or citizenship ties to the USA, creating a false impression of legitimacy (8).

Additionally, predatory journals lack editorial team diversity, particularly those claiming to be international publishers. Naturally, an international board or editorial team has verified and certified scientists from many nations and geographical regions. In contrast, the composition of editors is limited to a specific location or continent. Another feature is that the boards of predatory journals lack enough personnel. Typically, the committees will have two or three editors as their complete editorial team (Grudniewicz et al. 2019).

Moreover, predatory journals create fictional and fake board members with counterfeit credentials to boost the number of members. In other circumstances, the journals illegally include famous scientists without obtaining consent. For instance, a predatory publisher of open-access academic journals (OMICS) published an article, "Advances in Pharmacoepidemiology & Drug Safety," in 2012. OMICS claimed that two chief editors edited the paper. One editor from Pittsburgh University, Robert H. Howland, distanced himself from the company in 2016, stating that OMICS listed him as a chief editor without his consent. Predatory journals also use the photos of famous publishers without obtaining permission (Masic 2017).

Predatory publishers also have websites that are often poorly maintained compared to other legitimate publishers. A common feature seen includes the use of dead links, webpages having contents that include misspellings and grammatical mistakes, and plagiarized images drawn from the Internet. Another feature is republishing using alternate outlets without mentioning the original writers. Predatory publishers rebrand such articles as original work. A perfect example is the case of OMICS, which rebranded many previously published journals using subsidiary publishers like Hilaris, located in Belgium. When the activities of such publishers are discovered, they remove their logos from previous papers only to republish them in the same journals as the original work. For instance, when OMICS came under scrutiny, the publishers removed their logo from the journals only to republish them as purported original work (Dadkhah et al. 2016).

Aside from fake websites, editors, and locations, predatory publishers are skilled at forging digital object identifiers (DOIs). DOIs are unique identification tags generated by designated companies that aid in cross-referencing and retrieving papers. One of the largest sources of DOIs is Crossref, a nonprofit company. After the Federal Trade Commission (FTC) ruling against OMICS in 2019, Crossref ceased providing OMICS with DOIs to curb predatory activities. However, OMICS allegedly started generating fake and random DOIs to accompany all publications and research papers. Consequently, unsuspecting publishers can easily fall prey to OMICS, thinking the publisher adheres to all publishing standards, including DOIs (Masic 2017).

These behaviors reflect Beall's original definition of predatory publishers: entities exploiting the author-pays model by deceiving researchers and failing to uphold legitimate editorial standards (Beall 2017). Although Beall's list was discontinued in 2017, his evaluative framework remains foundational and is still used by updated exclusion lists and quality-check tools (Beall 2017).

7.4 The Range of Predatory Journals and the Caveats of Beall's Criteria

Although Beall described the features shared with most predatory publishers, some legitimate publishers were incorrectly labeled as predatory due to overlap. For instance, some journals may have appeared predatory due to editing errors that were overlooked. Moreover, some legitimate businesses fall victim to predatory brands due to their association with predatory publishers. Predatory journals apply measures to redirect legitimate journals, including taking over accidentally lapsed domains or using fake copy websites where original papers are displayed. The use of another journal's domain or website is now known as hijacked and clone journals (Jalalian and Mahboobi 2014). In addition to the above-mentioned features, some predatory publishers list scientists in their editorial boards without their consent (Nelson and Huffman 2015).

Predatory publishers also have enough capital to acquire legitimate journals. For instance, the OMICS publishing company managed to obtain two legitimate Canadian journals. In such a case, all legitimate articles published by the journal risk being labeled substandard. Since Crossref blocked OMICS, all legitimate journals published before the bans are ignored (Masic 2017). In addition, some legitimate journals from lower-income countries have appeared among the listed predatory journals. Poor publishing efforts put such journals at risk of inclusion in Beall's list. Since resource-poor institutions try to cut operating costs and subscriptions, the journal sites appear unprofessional and unreliable. Some local journals and university publications from developing countries are among the most at-risk journals (Demir 2018).

Beall, being one of the first to compile a list of predatory journals, holds significant influence. However, it is important to note that his list may reflect his personal views and attitudes toward certain publishers. From numerous interviews where Beall commented on predatory journals, prejudice and disdain toward open-access publishing were evident. On the other hand, Beall demonstrated a preference for subscription-based publishers. Such an attitude predisposed the unfair listing of legitimate journals due to their association with features he considered suspicious (da Silva and Tsigaris 2018).

7.5 Factors Influencing the Sprouting and Sustenance of Predatory Publishing

Several factors have created a fertile ground for the thriving of predatory publishers over the past ten years. First, the introduction and adoption of digitalization of legitimate publishing created a loophole for predatory publishers. Nevertheless, it enabled everyone with Internet access to view journals and has helped cut down on printing and distribution costs, making it easier for the publishing industry to flourish. With digitalization, a novel model of publishing has cropped up. Free access enabled a mechanism to collect revenue for accessing digital content in several ways besides subscription fees. As open-access journals became prominent, article processing charges (APCs) became popular. APCs are premised on the willingness of authors to pay for certain publishing services. Since publishers gain financially from APCs and subscriptions, predatory journals have continued to exist. Moreover, research funding, particularly in the biomedical field, is a common source of income and funding for predatory publishers. Individual funding research does not select the publishers for the scientists, thus creating a loophole for dubious publishers (Björk et al. 2020; Pyne 2017).

In the academic publishing industry, the quantity of publications offers an advantage, sometimes compromising quality. Moreover, most scientific accolades are awarded according to publication-based productivity. Besides the number of publications, productivity analysis focuses on the number of citations, the awarded grants per journal, the rank, and the authorship order assigned to any researcher. The quantity reward system tempts many researchers to use dubious methods to increase perceived publication volume. Consequently, some researchers direct their efforts toward quantity versus quality of research work. Moreover, where applicable, researchers split the findings of a single paper into multiple publications to create an impression of high productivity, a concept termed "salami slicing" (Sharma and Verma 2018). Vitek Tracz remarked on the speed with which publications came out at a rate greater than readers could consume, driven by the need for acknowledgment and rewards (Tracz 2012). Since most researchers become aware of predatory publishers' availability, they often overlook the standards required for any publication. Predatory journals then thrive on the researchers' desires for more publication volume (Demir 2018).

One epidemiologist from Stanford University noted the tendency of researchers to publish solely for promotion and tenure purposes. Most faculty appraisals require members to publish frequently to maintain status, thus tempting some members to opt for predatory publishers who only aim to create the most profit from academicians. For instance, in 2010, in India, the University Grants Committee (UGC) introduced Academic Performance Indicators (API). According to the API, every faculty member and Ph.D. candidate must achieve a certain minimum number of publications to succeed and progress academically. In 2013, India had the most significant number of authored journals published by predatory publishers, with 35% of such journals globally. Moreover, India had the most significant percentage (27%) of predatory journals (Demir 2018).

Some of the conditions set up by the UGC in India created an environment for predatory journals. For instance, even teaching-focused faculty with minimal experience in research have also been required to publish. Moreover, the UGC created a framework for awarding higher points for journals published abroad or by international companies relative to locally published documents. It is essential to note that the acceptance rate of papers presented by researchers from low- and middle-income countries is relatively low. Consequently, predatory journals that brand themselves as international publishers can easily exploit Indian researchers. The reward committees in Nigeria also require researchers to publish in international journals. As expected, Nigeria also has a higher number of journals published by predatory journals (Shrestha et al. 2019).

In other countries, like Italy, the national awarding systems reward researchers for publishing in specified journals. High-impact journals and those listed on the Web of Science are examples of high-impact journals. Some predatory journals are indexed under Scopus, while others falsely claim to be affiliated with Scopus. In one study on professors looking forward to promotions, 5% of the professors unknowingly published in predatory journals. Of note is that these professors published in low-impact journals, thereby increasing the risk of falling prey to predatory journals (Rahman et al. 2014).

7.6 Reasons to Be Concerned About Publishing in Predatory Journals

The publishing industry is trust based. The researchers trust the publishers to carry out advertised work, such as peer review. The editors trust the genuineness of the researchers in coming up with research findings. The scientific community also trusts publishers to do thorough evaluations of submitted articles to maintain scholarly standards. Finally, publishers also trust the genuine intent of clients to pay the fees. A breach in any of the trust chains results in tarnishing the research industry. Relying on the principle of trust, researchers can easily fall victim to predatory publishers who manipulate credentials to exploit researchers' trust and occasional naivety. Consequently, some legitimate research work ends up in the blacklisted journals, discouraging future scholarly engagement for some researchers (Grudniewicz et al. 2019; Cobey et al. 2018).

Fortunately, experienced researchers have the trained eye to identify predatory journals either through anecdotal accounts or through Beall's criteria. However, little or no training on publishing standards and ethics takes place in most institutions. In contrast, early career researchers are expected to start publishing and, as such, can easily fall prey to predatory publishers if left unguided (Beall 2012).

Most high-impact journals come from high-income countries with English as their native language. High-income countries benefit from the structural advantage of economies of scale, particularly in printing documents. Moreover, the costs associated with gaining access to subscription-based journals impede researchers from low-income countries from gaining access and building on current research work. Leaving out researchers from low-income countries creates a snowball effect in the acceptability of their manuscripts. First, limited access to high-end research implies that the researchers deal with low-end questions. Consequently, the high-impact journals display minimal interest in the purported inferior topics, creating and perpetuating a vicious cycle (Eriksson and Helgesson 2017). PubMed, an open-access publisher, allows all research funded by the National Institute of Health (NIH) to be published in its open-access archive. Since NIH funds researchers from several countries, low-income states included, such publications may appear and be listed as predatory. Adequate research must therefore be conducted to claim that predatory journals corrupt or pollute existing scientific databases (Goldacre 2017).

Regardless, predatory journals create a threat to scientific integrity. They have the potential to contaminate data in the scientific field (Shamseer et al. 2017). While scientists are discouraged from reading, citing, or publishing in predatory journals, this can only be implemented in full if the list is known to the researchers (Beall 2016). Four scenarios explain the potential inclusion of some predatory journals on approved sites: the journal may be relics from before vetting, have been previously blacklisted but later met the minimum standards, have been false positives according to Beall's criteria, or appear predatory despite being listed in PubMed (Beall 2016; Manca et al. 2018).

Another concept relating to compromising the integrity of science is that predatory publishers prevent credible information from reaching legitimate scientific databases. Most scientists assume that all information published in predatory journals is false. However, some unsuspecting researchers publish legitimate findings and results that end up being wrongfully blocked (Cobey et al. 2019). A framework for analyzing such data must be established, including investigating the origin, the authors, the legitimacy of the tools applied in collecting and analyzing data, and the legitimacy of the produced results and explanations (Cobey et al. 2019). The overall assumption is that research published in verified journals is more error-free than that published in predatory sources. However, research has shown errors and distortions in some publications, even those considered legitimate publishers (Ioannidis 2005). In 2015, John Ioannidis, an epidemiologist at Stanford University, published a study questioning the truthfulness of most published works. John wanted to expose why most of the published research may be flawed and how unchecked studies circulating in academic papers contribute significantly to the body of evidence (Ioannidis 2005). Although further research that corrects such data is considered natural scientific evolution, such corrections do not always occur. As a result, circulating facades are accepted as true (Fanelli 2013). Contrary to popular belief, unnoticed research in blacklisted journals may be the only source of counter-information. As a result, drawing a clear line between perceived reality and the source of publication deprives the scientific field of critical research (Shamseer et al. 2017).

Another valid reason for considering predatory journals is that they act as an indictment of legitimate journals. One is surprised why predatory journals attract such overwhelming submissions. The data suggests that a thin differentiating factor exists between legitimate and predatory publishers. The lack of transparency and integrity, particularly in legitimate journals, such as biases and prejudice toward low-income researchers, could have contributed to the surge of predatory publishers (Demir 2018). Moreover, one might ask why people opt to make submissions to poorly operating publishers while ignoring legitimate publishers. Perhaps publishers need to focus on structural inequalities and create an equal ground for all researchers to publish in high-impact journals (Alberts et al. 2014).

7.7 Ethical Standards for Publications

Five ethical standards cut across all publications and serve as the guiding principles for all publications. The first ethic entails complete respect for intellectual property and making every effort to avoid plagiarism of original work. All borrowed ideas must be referenced both in the text and in the bibliography sections as per this standard. Guidelines exist for the utilization of direct quotes according to the several referencing styles used in publications. Plagiarism and lack of respect for intellectual property can lead to litigation for copyright infringement and being blocked from reputable scientific journals (Committee on Publication Ethics [COPE] 2019).

All copyrighted materials, including images, graphs, and statistics, must be referenced appropriately. Most predatory publishers fail to adhere to this basic principle of respect for intellectual property (Beall 2016).

The second ethical principle is absolute respect for human dignity in research. All results must be made publicly available with the appropriate ethical approval if human participants were involved after completing the paper. Whether the results are positive, negative, or inconclusive, they must be put in the public domain, mainly when dealing with human subjects. Moreover, the researchers must endeavor to declare any conflict of interest while carrying out the research and the source of funding for the study (World Medical Association 2013).

The third principle is intentionally avoiding the development of a conflict of interest. It is prudent for researchers to prevent such conflicts, particularly financial conflicts of interest. In the event of one, the conflict must be declared. Moreover, if a paper is free of such conflicts, the researcher must indicate that no such conflict exists (International Committee of Medical Journal Editors [ICMJE] 2019).

The fourth principle entails carrying out legal action on the manipulation of citations. Some individuals randomly cite journals to increase their citation index, which raises the credibility of publications. However, such a practice is highly discouraged as it risks legal action and the rejection of the entire research work (Committee on Publication Ethics [COPE] 2019). Finally, all publishers and researchers must avoid research misconduct at all costs. Research misconduct entails falsifications, plagiarism, and other illegalities within the publishing business. All editors are advised to reject suspicious submissions and present them before the necessary authorities to avoid attracting legal implications and litigation (Committee on Publication Ethics [COPE] 2019).

References

Abalkina A. Detecting a network of hijacked journals by its archive. arXiv. 2021a; https://arxiv.org/abs/2101.00000.

Abalkina A. Detecting a network of hijacked journals by its archive. Scientometrics. 2021b;126(6):1–26. https://doi.org/10.1007/s11192-021-03879-1.

Alberts B, Kirschner MW, Tilghman S, Varmus H. Rescuing US biomedical research from its systemic flaws. Proc Natl Acad Sci. 2014;111(16):5773–7. https://doi.org/10.1073/pnas.1404402111.

Beall J. Predatory publishers are corrupting open access. Nature. 2012;489(7415):179. https://doi.org/10.1038/489179a.

Beall J. Criteria for determining predatory open-access publishers. Scholarly Open Access; 2015. https://beallslist.net/criteria.

Beall J. Predatory journals: ban predators from the scientific record. Nature. 2016;534(7607):326. https://doi.org/10.1038/534326a.

Beall J. Why Beall's list went dark; Beall maintained 56 point criteria. Retraction Watch; 2017. https://retractionwatch.com/2017/01/17/bealls-list-publisher-blacklist-removed/.

Björk BC, Kanto-Karvonen S, Harviainen JT. How frequently are articles in predatory open access journals cited? Publica. 2020;8(2):17. https://doi.org/10.3390/publications8020017.

Bohannon J. Who's afraid of peer review? Science. 2013;342(6154):60–5. https://doi.org/10.1126/science.342.6154.60.

Cobey KD, Lalu MM, Skidmore B, Ahmadzai N, Grudniewicz A, Moher D. What is a predatory journal? A scoping review. F1000Res. 2018;7:1001. https://doi.org/10.12688/f1000research.15256.2.

Cobey KD, Grudniewicz A, Lalu MM, Rice DB, Raffoul H, Moher D, Al-Khatib A. Knowledge and motivations of researchers publishing in presumed predatory journals: a survey. BMJ Open. 2019;9(3):e026516. https://doi.org/10.1136/bmjopen-2018-026516.

Committee on Publication Ethics. Code of conduct and best practice guidelines for journal editors. COPE; 2011. https://publicationethics.org/resources/code-conduct.

Committee on Publication Ethics. Core practices. COPE; 2019. https://publicationethics.org/core-practices.

da Silva JAT, Tsigaris P. What value do journal whitelists and blacklists have in academia? J Acad Librariansh. 2018;44(6):781–92. https://doi.org/10.1016/j.acalib.2018.09.017.

Dadkhah M, Maliszewski T, Jazi MD. Characteristics of hijacked journals and predatory publishers: our observations in the academic world. Trends Pharmacol Sci. 2016;37(6):415–8. https://doi.org/10.1016/j.tips.2016.04.002.

Davis P. SCIgen and the misconduct in science: a hoax paper story. The Scholarly Kitchen; 2010. https://scholarlykitchen.sspnet.org/2010/01/05/scigen-and-the-misconduct-in-science.

Demir SB. Predatory journals: who publishes in them and why? J Informet. 2018;12(4):1296–311. https://doi.org/10.1016/j.joi.2018.10.008.

Editage Insights. Hijacked journals vs predatory journals; Abalkina on COVID-19 hijacks. 2021. https://www.editage.com/insights.

Eriksson S, Helgesson G. The false academy: predatory publishing in science and bioethics. Med Health Care Philos. 2017;20(2):163–70. https://doi.org/10.1007/s11019-016-9740-3.

Eysenbach G. Peer-review and fraud detection in open access publishing. J Med Internet Res. 2008;10(2):e13. https://doi.org/10.2196/jmir.1065.

Fanelli D. Redefine misconduct as distorted reporting. Nature. 2013;494(7436):149. https://doi.org/10.1038/494149a.

Goldacre B. The WHO joint statement from funders on trials transparency. BMJ. 2017;357:j2816. https://doi.org/10.1136/bmj.j2816.

Grudniewicz A, Moher D, Cobey KD, Bryson GL, Cukier S, Allen K, et al. Predatory journals: no definition, no defence. Nature. 2019;576(7786):210–2. https://doi.org/10.1038/d41586-019-03759-y.

International Committee of Medical Journal Editors. Recommendations for the conduct, reporting, editing, and publication of scholarly work in medical journals. ICMJE; 2019. http://www.icmje.org/recommendations/.

Ioannidis JPA. Why most published research findings are false. PLoS Med. 2005;2(8):e124. https://doi.org/10.1371/journal.pmed.0020124.

Jalalian M, Mahboobi H. Hijacked journals and predatory publishers: is there a need to re-think how to assess the quality of academic research? Walailak J Sci Technol. 2014;11(5):389–94.

Manca A, Moher D, Cugusi L, Dvir Z, Deriu F, Folli F. How predatory journals leak into PubMed. Can J Cardiol. 2018;34(11):1461–5. https://doi.org/10.1016/j.cjca.2018.07.007.

Masic I. Predatory publishing—experience with OMICS international. Med Archiv. 2017;71(5):304–7. https://doi.org/10.5455/medarh.2017.71.304-307.

Nelson N, Huffman J. Predatory journals in library databases: how much should we worry? Ser Libr. 2015;69(2):169–92. https://doi.org/10.1080/0361526X.2015.1080782.

Poynder R. Open and shut? In: The open access interviews: Richard Poynder on scholarly publishing; 2008. Open and Shut. https://poynder.blogspot.com.

Pyne D. The rewards of predatory publications at a small business school. J Sch Publ. 2017;48(3):137–60. https://doi.org/10.3138/jsp.48.3.137.

Rahman AIMJ, Dexters N, Engels TC. Predatory open access journals in a performance-based funding model: common journals in Beall's list and in the VABB-SHW. J Assoc Inf Sci Technol. 2014;66(2):319–27. https://doi.org/10.1002/asi.23165.

Research and Markets. Global scientific publishing market report 2021–2026. Research and Markets; 2021. https://www.researchandmarkets.com/reports/5147833/global-scientific-publishing-market-report-2021-2026

Shamseer L, Moher D, Maduekwe O, Turner L, Barbour V, Burch R, et al. Potential predatory and legitimate biomedical journals: can you tell the difference? A cross-sectional comparison. BMC Med. 2017;15:28. https://doi.org/10.1186/s12916-017-0785-9.

Sharma H, Verma S. Predatory journals: the rise of worthless biomedical science. J Postgrad Med. 2018;64(4):226–9. https://doi.org/10.4103/jpgm.JPGM_316_18.

Shrestha J, Subedi S, Shokati B, Chaudhary A. Predatory journals: a threat to scholarly publishing. J Edu Res. 2019;8(1):89–101. https://doi.org/10.3126/jer.v8i1.28764.

Tracz V. Publish or perish: the pace of scientific publishing and its consequences [Conference presentation]. SlideShare; 2012. https://www.slideshare.net/vitektracz/publish-or-perish-the-pace-of-scientific-publishing.

Van Noorden R. Computer-generated nonsense paper accepted for conference. Nature News. 2014; https://www.nature.com/news/computer-generated-nonsense-paper-accepted-for-conference-1.14763.

Ware M, Mabe M. The STM report: an overview of scientific and scholarly publishing. 6th ed. International Association of Scientific, Technical and Medical Publishers; 2021.

World Medical Association. Declaration of Helsinki: ethical principles for medical research involving human subjects. JAMA. 2013;310(20):2191–4. https://doi.org/10.1001/jama.2013.281053.

8 Strategies to Tackle Academic Misconduct in Academic Writing

Özgür Çelik and Temel Serdar Yılmaz

8.1 Introduction

Academic integrity lies at the very heart of academic writing, requiring students, faculty, and researchers to maintain *honesty, trust, fairness, respect, responsibility, and courage* in their pursuit of knowledge. This enables their research endeavors to be credible and valid and support trustworthy and respectful collaboration among all stakeholders of academia (ICAI 2021). In recent years, however, there has been a growing global concern of academic misconduct in academic writing.

Studies have indicated the prevalence of academic misconduct in various disciplines and contexts (e.g., Dhingra and Mishra 2014; Fanelli 2009; Feenstra et al. 2021; Felaefel et al. 2018; Okonta and Rossouw 2013). The most reported types of academic misconduct in these studies are data fabrication and falsification, gift or ghost authorship, plagiarism, and violation of research ethics regulations. Collusion and contract cheating are also other prevalent types of academic misconduct committed by students (Clarke and Lancaster 2006; Lancaster and Clarke 2012; McCabe 2005; Trost 2009).

Though such incidences of misconduct may be viewed as personal breaches of academic integrity at first glance, their impact transcends individual projects and affects all academia. They induce inaccurate research results, which may cause distortion of scientific knowledge, erode the reputations of academic institutions and research, create mistrust among academics and diminish scientific collaboration, mislead policymakers and decision-makers, resulting in loss of time, financial resources, and labor of force (Martinson et al. 2006; Steneck 2006).

Ö. Çelik (✉) · T. S. Yılmaz
School of Foreign Languages, Balıkesir University, Balıkesir, Türkiye

A. K. Abu-Shaheen et al. (eds.), *Introduction to Research Ethics and Academic Integrity*, https://doi.org/10.1007/978-981-96-9678-9_8

Among the factors exacerbating academic misconduct occurrences, the most prominent are lack of training in ethics and integrity (El-Shinawi et al. 2016; Felaefel et al. 2018) and so insufficient knowledge of researchers and students in academic integrity (e.g., El-Shinawi et al. 2016), lack of clear and consistent institutional policies and supervision (Heitman and Litewka 2011), insufficient academic writing skills in English (Cameron et al. 2012; Heitman and Litewka 2011; Pecorari 2016; Vasconcelos et al. 2009), varying cultural perceptions in authoritative source, rights of intellectual property, and what constitutes plagiarism (Heitman and Litewka 2011).

Concerning the severity of prevalence and consequences of academic misconduct in scholarly writing along with the diversity and multiplicity of factors intensifying this situation, there is an urgent need for a multilevel approach to detect, deter, and mitigate all forms of breaches of academic integrity. The purpose of this study, therefore, is to identify and evaluate the promising strategies to cope with academic misconduct in academic writing using Kenny and Eaton's (2022) 4 M framework comprehensively and systematically.

8.2 Method

In this study, we followed a systematic process to identify, refine, and categorize strategies that can be adopted to mitigate academic misconduct in academic writing. We started with a review of relevant literature to identify potential strategies for addressing academic misconduct in academic writing. We conducted a literature search on Google Scholar using specific keywords related to academic integrity, academic misconduct, and academic writing. In the end, we found 61 studies that were considered potential sources for our analysis. Subsequently, we delved into each study and extracted strategies mentioned for mitigating academic misconduct. To do this, we carefully read through each study, made note of all the mentioned strategies, and collected all relevant strategies identified. At the end of this process, we identified a total of 252 strategies in our literature database.

After identifying 252 strategies, we realized that many of them were similar or overlapping. Therefore, we removed duplicates and merged similar ones to refine the strategy list. The objective was to ensure that each strategy was unique. As a result, we were able to distill the strategies into a more refined and unique list. After refining the strategy list, using the 4 M framework (Kenny and Eaton 2022), we categorized them into four levels as micro-, meso-, macro-, and mega-level strategies. After refining and categorizing the strategies, we conducted an analysis to identify overarching strategy themes. We examined the strategies in each category to find similarities and underlying principles that could be grouped into broader themes. This helped us summarize strategies into broader themes and offer insights on mitigating academic misconduct in writing.

8.3 Micro-, Meso-, Macro- and Mega-level Strategies

Maintaining academic integrity in all scholarly works is essential for educational excellence and scholarly advancement. It is crucial to embrace a holistic and hierarchical approach to establish a culture of academic integrity that moves beyond individuals to society at large. Therefore, it is essential to draw a framework for strategies to be adopted at micro-, meso-, macro-, and mega-levels. This section encapsulates the strategies that can be adopted to mitigate academic misconduct in academic writing.

8.4 Micro-level Strategies

Individuals are the building blocks of institutions and societies. Therefore, it is crucial to focus on individual actions and interventions that cultivate a culture of integrity and ethical behavior. In this section, five micro-level strategies derived from the literature are presented.

8.4.1 Educating Individuals Through Workshops or Training

Delivering well-designed training sessions and educational workshops on academic writing promises a strong foundation for equipping students with necessary knowledge and skills. Such programs help students learn the foundational and practical aspects of academic integrity (Ard and Ard 2019), enhance understanding and knowledge of academic integrity, reveal the gaps in students' knowledge about academic integrity (Curtis et al. 2022), encourage students to take the ownership of their own learning (Lysiak 2020), and help mitigate plagiarism cases among students (Nikjo et al. 2021). Whether delivered online or in person, such educational interventions propose a strong opportunity to proactively approach academic misconduct by cultivating a shared understanding of academic integrity in individuals.

8.4.2 Fostering Personal Development and Mentorship

Research shows that the development of academic integrity values is strongly correlated with personal integrity values (Hussein 2017). Therefore, to foster a personal commitment to academic integrity values, it is essential to encourage students to self-reflect on ethical values. Rather than just teaching the mechanics aspects of academic writing, framing the issue from awareness and development perspective helps students to become a part of the community of scholars (McGowan 2005). This can be best achieved by creating a learning environment where students gain the ability to contemplate the outcomes of their actions and behave accordingly. Pairing students with experienced mentors can significantly enhance the individual commitment to academic integrity through fostering personal development (Khan et al. 2020b). Inadequate mentorship is one of the reasons for the rise of academic misconduct (Ochroch and Eckenhoff 2011). Therefore, it is essential to foster personal development of students through effective mentorship as a proactive attempt to mitigate academic misconduct incidents and contribute to the culture of academic integrity.

8.4.3 Familiarizing Students with Technology Use

Technology can be considered as a threat and an opportunity at the same time for academic integrity. Especially after the emergency remote teaching period, the use of technology in academic settings raised concerns in terms of academic integrity (Solmon 2018). Moreover, the rise of artificial intelligence spurred the discussion of whether such technologies are a threat or asset to academic integrity (Rodrigues et al. 2024). However, responsible and effective use of technology offers significant outcomes in maintaining academic integrity. Research shows that technology-based interventions significantly contribute to the climate of academic integrity through familiarizing students with the correct use of such technologies and improving their knowledge and attitudes (Cronan et al. 2017; Hanbidge et al. 2020). It is crucial to note that the use of technology for academic purposes, especially in academic writing, is a skill that should be developed through the principles of academic integrity.

8.4.4 Understanding Student Perceptions and Attitudes

Exploring student perceptions, attitudes, needs, and behaviors about academic integrity is essential and many studies set out to explore these constructs (e.g., Allen et al. 1998; Elzubeir and Rizk 2003; Kumar et al. 2023; MacLeod and Eaton 2020; Robinson and Glanzer 2017). These studies provide valuable insights into understanding the underlying reasons for academic misconduct and developing proactive measures to mitigate such incidents. Moreover, changing behaviors should start with understanding the roots of behaviors. Therefore, it is a fruitful endeavor to explore student attitudes and perceptions to develop a more targeted intervention programs to maintain academic integrity.

8.4.5 The Earlier, the Better

Writing development is a process where students start to develop their authorial identity. They develop their own writing style and habits. It can be seen as a habit-formation process in terms of writing. Lack of training may result in flawed writing habits, and this can be carried to the later stages of education (Bruton and Childers 2016). Therefore, it is necessary to start educating students about academic integrity as early as possible in school settings (Wan and Scott 2016). Early and intensive exposure to the principles of academic integrity can reinforce ethical behavior. Therefore, it is recommended that students should be exposed to academic integrity even before they step into the school (Stephens 2016).

These micro-level strategies strive to expand individuals' capacity to embrace and internalize the principles of academic integrity in their writing process. Although these strategies promise a strong foundation for students to be responsible writers in their academic works, the success of these strategies requires a holistic and multi-stakeholder approach within institutions.

8.5 Meso-level Strategies

Meso-level strategies strive to promote academic integrity through departmental interventions. Such strategies require the participation of various stakeholders and systemic collaboration. The primary aim of these strategies is to create an environment that fosters an academic integrity culture.

8.5.1 Delivering Faculty-Wide Training

Academic integrity was regarded as a student issue for a long time (D. McCabe and Pavela 2000), and most of the interventions were aimed at students only. However, it was then seen that academic integrity requires a multi-stakeholder responsibility, not just a student issue (Bretag et al. 2011). The evidence shows that the training of stakeholders offers vital implications for fostering a culture of academic integrity (Khan 2024). Therefore, providing training sessions to all stakeholders of educational processes, especially those involved in writing pedagogy (i.e., students, teachers, supervisors, and librarians), can facilitate a better learning environment for students in their writing journey. Moreover, the nature of these training sessions should address beyond teaching just mechanical aspects of writing with integrity but focus more on ethical and rhetorical aspects of academic writing ethically (Howard and Jamieson 2021). Also, it should be ensured that the training of faculty is well-planned, continuous, and it should meet the needs of the stakeholders.

8.5.2 Integrating Academic Integrity into Syllabi

Approach to academic integrity, especially writing with integrity in general and plagiarism in particular, has undergone certain paradigm shifts. Earlier approaches saw integrity as the responsibility of students who are habituated to cheating (East 2009; McCabe and Pavela 2000), and teachers were responsible for dealing with student misconduct (Paine 1994). Later approaches adopted a more educational perspective and saw integrity as a teaching and learning issue (Bertram Gallant 2017). Howard and Jamieson (2021) also highlighted the importance of teaching the ethics and rhetoric of academic writing rather than just teaching the mechanical aspects of it. The new approach to academic integrity revealed the necessity of integrating the values, principles, and messages of academic integrity into the course syllabi. This integration should not only focus on the detection and reaction of academic misconduct but also cover the deterrence of academic misconduct through educational and proactive interventions.

8.5.3 Developing and Enforcing Departmental Policies and Procedures

It is widely accepted that developing and enforcing effective academic integrity policies is a significant facilitator of fostering a culture of academic integrity (Morris 2016). Departmental academic integrity policies and procedures ensure that all stakeholders share the same understanding, expectations, and consequences related to academic integrity. The lack of a shared understanding among students and faculty may undermine the intended academic integrity outcomes (Spain and Robles 2011). Therefore, it is critical to craft policies and procedures that comply with the school culture and audience. These policies provide a basis for creating an environment that supports students to be responsible writers who develop their writing skills around the principles of academic integrity. Moreover, by emphasizing the importance of owning their own work, these policies and procedures encourage students to develop critical thinking and research skills needed for their academic progression.

8.5.4 Aligning Assessment with Academic Integrity

An assessment design that carries the elements of a personalized and unique assessment is an effective way to discourage students from committing academic misconduct (Egan 2018; Hamilton and Richardson 2007). Especially for academic writing, authentic assessment provides clear opportunities to reach intended outcomes through formative processes (Morris 2016). However, the study of Ellis et al. (2020) shows that authentic assessment may not always assure academic integrity, especially in terms of contract cheating. Their study analyzes a large number of assignments based on five authenticity factors (frequency, fidelity, complexity, real-world impact, and feed-forward) and finds that the outsourcing of the assignments does not depend on the authenticity level of the assignment, suggesting that what authentic assignment means should be redefined or reframed. On the other hand, the study of Sotiriadou et al. (2020) shows that scaffolded assessment promises to be a good strategy to prevent academic misconduct. Openo (2024) proposes some strategies to align assessment with academic integrity. These strategies include designing tasks that are relevant and simulate real-word activities; encouraging collaboration and reflection; leveraging the tasks degree of authenticity; tailoring assessment to specific contexts; allowing for iterative feedback; and relating the tasks to the skills and processes found in the professional fields. With all these in mind, it can be argued that carefully designed assessments that align with academic integrity principles help mitigate academic misconduct and create an environment where learners develop their writing skills in a responsible and ethical way.

To summarize, meso-level strategies provide a strong foundation for promoting academic integrity within institutional practices and fostering an ethical learning environment. These strategies cover departmental policies, curriculum, training of faculty members, and assessment. Effective implementation of these strategies helps create a strong framework for ethical academic standards within institutions. However, their effectiveness depends on the commitment and collaboration of every member of the academic community.

8.6 Macro-level Strategies

Macro-level strategies are crucial in fostering a culture of academic integrity and addressing academic misconduct at the institutional level. These strategies provide a comprehensive framework that supports and enforces academic integrity across all institutional bodies, requiring the commitment and coordination of all the stakeholders, from top management to each individual.

8.6.1 Providing an Institutional Policy Framework

Institutional policy frameworks serve as a roadmap for departmental policies by establishing a baseline for expectations for all the members of the academic community regarding ethical academic conduct. Such frameworks serve as a foundational reference point for meso-level policies. Therefore, before developing the meso-level policies, institutions should work on developing macro-level policy framework to ensure consistency, fairness, and transparency across the institution. However, the development of these frameworks requires robust and educative approaches and should be aligned with quality assurance standards (Glendinning 2022). An approach without intentional and systematic attention and that focuses on policing and punishing academic integrity has proven to be ineffective in creating an ethical learning environment (Bertram Gallant and Drinan 2006). It should also be noted that such policy frameworks are only effective when they originate from the school culture, not imposed on them (Whitley and Keith-Spiegel 2001). In this perspective, Bertram Gallant and Drinan (2008) propose a four-step institutionalization model for academic integrity, which comprises recognizing and commitment to academic integrity at all levels of the institution, generating systemic responses, implementing responses in an educative way, and finally, institutionalization. Within this scope, institutional academic integrity policy frameworks play a pivotal role in shaping the academic writing developments of students by establishing clear expectations, providing necessary guidance, and fostering an environment conducive to ethical scholarship.

8.6.2 Running Institution-Wide Awareness Campaigns

Running awareness campaigns is essential for increasing student engagement and awareness on academic integrity-related topics (Khan et al. 2020a). These campaigns play a crucial role in educating the academic community members about the fundamental values of academic integrity and ultimately help create institutional ethos. Studies also show that these campaigns effectively raise student awareness and help mitigate academic misconduct (El-Shinawi et al. 2016). Moreover, institution-wide academic integrity campaigns provide clear implications for students' academic writing development by emphasizing the importance of ethical writing practices such as ethical authorship, proper citation, originality, etc. Ultimately, academic integrity campaigns are effective tools to nurture students' growth as ethical and proficient writers to enhance the quality of their academic work. However, to maximize the effectiveness of such campaigns, it is essential to adopt creative approaches to engage all members of the academic community through different channels of communication in a systematic and continuous way.

8.6.3 Providing Institution-Wide Support Services

Fostering a culture of academic integrity is strongly correlated to the degree of providing support for all academic community members at the policy level (Bretag et al. 2011). Conversely, the lack of support may lead to inconsistencies in enacting the values of academic integrity (Bertram Gallant and Drinan 2006; Kier and Ives 2022; Stoesz and Eaton 2020). Research also shows that students' perceptions of academic integrity are influenced by the level of support they get from their teachers (Laka and Paska 2023), and the inclusiveness of support services directly impacts their effectiveness (Davis 2022). As such, institution-wide support services provide significant implications for students' writing development by offering a range of resources and services, such as writing centers, libraries, online tutorials, and workshops. These support services, especially writing centers, offer guidance on avoiding plagiarism, properly citing sources, engaging critically with sources, and improving writing skills. In addition, writing centers make sure that students do not engage in academic misconduct by using a variety of strategies (Garwood 2022). Consequently, institutional support services foster a culture of integrity and actively contribute to developing competent and ethical writers.

To sum up, macro-level strategies are fundamental to tackling academic misconduct by creating a teaching and learning environment where students are encouraged to seek help, not seek unethical shortcuts due to the lack of support. Although these strategies are considered wide-ranging and effective, the success of macro-level strategies depends on dedication, cooperation, and the capacity to implement them effectively.

8.7 Mega-level Strategies

The ultimate purpose of mega-level strategies is to expand the initiatives to foster a culture of academic integrity and mitigate academic misconduct beyond individual institutions. These strategies are intended to target the wider educational communities, stakeholders, and societal frameworks. In the long run, megaDOUBLEHYPHENlevel strategies seek to contribute to ethical societal improvement.

8.7.1 Collaborating Between Institutions

Academic integrity is a universal value that is the outcome of shared understanding and collective strength. Therefore, mega-level approaches are necessary to extend the scope of academic integrity beyond individual institutions (Eaton et al. 2019). Collaboration between institutions serves as a versatile strategy to broaden the impact of best practices, thereby contributing to the development of unified standards for academic integrity. The different approaches to academic and research integrity among mega-level communities can worsen academic misconduct, including plagiarism and contract cheating, which provides easy options for those who engage in academic misconduct. In this respect, institutional collaboration to set a shared understanding and standards has a strong potential to mitigate academic misconduct and create a fair learning environment at the mega-level.

8.7.2 Engaging with External Stakeholders

It is well-established that an institutional culture of academic integrity requires the active participation of all relevant stakeholders in the academic community (East and Donnelly 2012; Kenny and Eaton 2022; Khan 2024). However, this multi-stakeholder approach should reach beyond the institution and cover mega-level stakeholders such as government bodies, industry partners, professional organizations, alumni, and even parents. Collaboration with mega-level stakeholders can enhance educational institutions' efforts to instill ethical principles and practices among students and faculty by providing diverse perspectives and resources. Various entities such as governmental education departments, quality assurance and accreditation bodies, professional academic associations, thematic academic networks, academic publishers, and educational content providers can be considered as influential mega-level stakeholders in creating a learning environment that fosters ethical academic writing development. These external stakeholders contribute to educational strategies and outcomes of the institutions.

In summary, mega-level strategies require collective and coordinated efforts across different sectors and levels of society to promote academic integrity. It is important to highlight that effective collaboration, involvement, and dedication are key factors in the success of mega-level strategies.

8.8 Conclusion

In this study, we set out to explore potential strategies that can be employed to mitigate academic misconduct in higher education academic writing. To do this, initially, we started with a comprehensive literature review and identified 252 strategies that have the potential to mitigate academic misconduct in academic writing. Subsequently, we refined the identified strategies and categorized them into four levels using the 4 M framework (Kenny and Eaton 2022), which includes micro-, meso-, macro-, and mega-levels. Refining helped us to analyze the strategies within a broader context and identify overarching strategy themes. This process resulted in a list of strategies that can address academic misconduct at different levels, from individual actions to societal-wide initiatives.

Micro-level strategies address elevating academic integrity at the individual level and play a critical role. These strategies cover providing mentorship for personal development and education through training, encouraging responsible use of technology, understanding student perceptions, and promoting early academic integrity education. Micro-level strategies are important because they equip individuals with the necessary knowledge, skills, and ethical frameworks to uphold integrity in learners' writing development. It can be suggested that focusing on micro-level strategies requires careful planning and promises significant outcomes. It is important to note that educators, institutions, and students must work together to implement these strategies (Hunter and Kier 2022), as academic integrity begins with the individual but is supported and enriched by the broader educational environment.

Meso-level strategies reflect the collaborative nature of academic integrity-related interventions, underscoring the significance of a joint effort between departments. Strategies such as faculty-wide training, integration of academic integrity messages into the curriculum, departmental policies, and an assessment design that complies with the principles of academic integrity highlight that academic integrity is beyond individual efforts. With a multi-stakeholder holistic approach (Morris and Carroll 2015), meso-level strategies help create a learning environment based on the principles of academic integrity. Such an environment where all members of the community share the same understanding and values in their particular department significantly reinforces the academic writing development of students by scaffolding their writing process in an ethical manner. However, institutional commitment is a strong determinant for the success of meso-level strategies.

Macro-level strategies emphasize the significance and necessity of institutional frameworks that act as a blueprint for leveraging a culture of academic integrity through an educational landscape. Macro-level strategies include a well-designed institutional framework, institution-wide academic integrity campaigns, and providing support services for educational community members. Moreover, these strategies underscore the need for a unified and systemic approach that promises a shared understanding from the top and to the bottom of an academic institution. However, it is worth noting that, besides being systemic, macro-level strategies should adopt an educative rather than a punitive approach (Bornsztejn 2022), suggesting that the focus is on leveraging academic integrity, not punishing academic misconduct. With these macro-level strategies, institutions should create an atmosphere where students are supported in their writing development, not punished due to their flawed writing habits.

Mega-level strategies highlight the importance of extending academic integrity beyond individual institutions to a broader educational and societal context. Mega-level strategies cover inter-institution collaboration and cooperation between academic institutions and related external stakeholders such as governmental bodies, professional organizations, industry partners or networks and associations. The importance of mega-level strategies lies in their potential to draw a strong authoritative framework for educational communities and promise a cohesive approach to upholding academic integrity. By fostering collaborations and cooperation among related stakeholders and educational institutions (Eaton et al. 2019), these strategies aim to create a robust framework for academic integrity that reflects shared values and collective efforts. Ultimately, mega-level strategies advocate for a societal-wide commitment to academic integrity, suggesting that the cultivation of ethical academic communities leads to ethical societal communities.

In conclusion, exploring micro-, meso-, macro-, and mega-level strategies highlights that a multi-faceted, holistic, and systemic approach to promoting academic integrity is required within the educational landscape. Each level of strategy contributes uniquely to developing a strong culture of integrity, which is essential for students' academic and professional growth, particularly for their academic writing development. This 4 M-strategy approach ensures that academic integrity is not just a student responsibility, and students are not passive recipients of academic integrity messages but active contributors to creating a culture of integrity that extends beyond the classroom.

Acknowledgments The authors acknowledge the use of AI tools during the preparation of this work. Grammarly, an AI-based language improvement tool, was employed to enhance the clarity and quality of the manuscript's language. Additionally, ChatGPT (Model 3.5) was utilized during the initial phases of the study for brainstorming purposes. The insights and conclusions presented in this work are solely those of the authors, who take full responsibility for the content and interpretations expressed herein.

References

Allen J, Fuller D, Luckett M. Academic integrity: behaviors, rates, and attitudes of business students toward cheating. J Mark Educ. 1998;20(1):41–52. https://doi.org/10.1177/027347539802000106.

Ard SE, Ard F. The library and the writing Centre build a workshop: exploring the impact of an asynchronous online academic integrity course. New Rev Acad Librariansh. 2019;25(2–4):218–43. https://doi.org/10.1080/13614533.2019.1644356.

Bertram Gallant T. Academic integrity as a teaching & learning issue: from theory to practice. Theory Pract. 2017;56(2):88–94. https://doi.org/10.1080/00405841.2017.1308173.

Bertram Gallant T, Drinan P. Institutionalizing academic integrity: administrator perceptions and institutional actions. NASPA J. 2006;43(4):61–81. https://doi.org/10.2202/1949-6605.1723.

Bertram Gallant T, Drinan P. Toward a model of academic integrity institutionalization: informing practice in postsecondary education. Can J High Educ. 2008;38(2):25–43. https://eric.ed.gov/?id=EJ815066.

Bornsztejn H. Do academic integrity policies within foundation studies programmes adopt an educative perspective for supporting students? J High Educ Policy Manag. 2022;44(5):428–42. https://doi.org/10.1080/1360080X.2022.2112272.

Bretag T, Mahmud S, Wallace M, Walker R, James C, Green M, East J, McGowan U, Patridge L. Core elements of exemplary academic integrity policy in Australian higher education. Int J Educ Integr. 2011;7(2):Article 2. https://doi.org/10.21913/IJEI.v7i2.759.

Bruton S, Childers D. The ethics and politics of policing plagiarism: a qualitative study of faculty views on student plagiarism and Turnitin ®. Assess Eval High Educ. 2016;41(2):316–30. https://doi.org/10.1080/02602938.2015.1008981.

Cameron C, Zhao H, McHugh MK. Perspective: publication ethics and the emerging scientific workforce: understanding "plagiarism" in a global context. Acad Med. 2012;87(1):51–4. https://doi.org/10.1097/ACM.0b013e31823aadc7.

Clarke R, Lancaster T. Eliminating the successor to plagiarism? Identifying the usage of contract cheating sites. In: Proceedings of 2nd international plagiarism conference; 2006. p. 1–13.

Cronan TP, McHaney R, Douglas DE, Mullins JK. Changing the academic integrity climate on campus using a technology-based intervention. Ethics Behav. 2017;27(2):89–105. https://doi.org/10.1080/10508422.2016.1161514.

Curtis GJ, Slade C, Bretag T, McNeill M. Developing and evaluating nationwide expert-delivered academic integrity workshops for the higher education sector in Australia. High Educ Res Develop. 2022;41(3):665–80. https://doi.org/10.1080/07294360.2021.1872057.

Davis M. Examining and improving inclusive practice in institutional academic integrity policies, procedures, teaching and support. Int J Educ Integr. 2022;18(1):14. https://doi.org/10.1007/s40979-022-00108-x.

Dhingra D, Mishra D. Publication misconduct among medical professionals in India. Indian J Med Ethics. 2014;11 https://doi.org/10.20529/IJME.2014.026.

East J. Aligning policy and practice: an approach to integrating academic integrity. J Acad Lang Learn. 2009;3(1):Article 1. https://journal.aall.org.au/index.php/jall/article/view/66.

East J, Donnelly L. Taking responsibility for academic integrity: a collaborative teaching and learning design. J Univ Teach Learn Pract. 2012;9(3) https://eric.ed.gov/?id=EJ996016.

Eaton SE, Chibry N, Toye MA, Rossi S. Interinstitutional perspectives on contract cheating: a qualitative narrative exploration from Canada. Int J Educ Integr. 2019;15(1):9. https://doi.org/10.1007/s40979-019-0046-0.

Egan A. Improving academic integrity through assessment design [monograph]. Dublin City University; 2018. https://doras.dcu.ie/22683/.

Ellis C, van Haeringen K, Harper R, Bretag T, Zucker I, McBride S, Rozenberg P, Newton P, Saddiqui S. Does authentic assessment assure academic integrity? Evidence from contract cheating data. High Edu Res Develop. 2020;39(3):454–69. https://doi.org/10.1080/07294360.2019.1680956.

El-Shinawi M, Mohamed KO, Fouad YA, Fahmy YM, Asar HA, Khalil MG, Anestidou L, El-Kamary SS, Mohamed MM. Assessing the awareness of Egyptian medical students about responsible conduct of research and research ethics: impact of an educational campaign. Account Res. 2016;23(4):199–218. https://doi.org/10.1080/08989621.2015.1127762.

Elzubeir MA, Rizk DEE. Exploring perceptions and attitudes of senior medical students and interns to academic integrity. Med Educ. 2003;37(7):589–96. https://doi.org/10.1046/j.1365-2923.2003.01552.x.

Fanelli D. How many scientists fabricate and falsify research? A systematic review and meta-analysis of survey data. PLoS One. 2009;4(5):e5738. https://doi.org/10.1371/journal.pone.0005738.

Feenstra RA, Delgado López-Cózar E, Pallarés-Domínguez D. Research misconduct in the fields of ethics and philosophy: researchers' perceptions in Spain. Sci Eng Ethics. 2021;27(1):1. https://doi.org/10.1007/s11948-021-00278-w.

Felaefel M, Salem M, Jaafar R, Jassim G, Edwards H, Rashid-Doubell F, Yousri R, Ali NM, Silverman H. A cross-sectional survey study to assess prevalence and attitudes regarding

research misconduct among investigators in the Middle East. J Acad Ethics. 2018;16(1):71–87. https://doi.org/10.1007/s10805-017-9295-9.
Garwood K. Supporting academic integrity in the writing Centre: perspectives of student consultants. In: Eaton SE, Christensen Hughes J, editors. Academic integrity in Canada: an enduring and essential challenge. Springer; 2022. p. 431–47. https://doi.org/10.1007/978-3-030-83255-1_22.
Glendinning I. Aligning academic quality and standards with academic integrity. In: Eaton SE, Curtis GJ, Stoesz BM, Clare J, Rundle K, Seeland J, editors. Contract cheating in higher education: global perspectives on theory, practice, and policy. Springer International Publishing; 2022. p. 199–218. https://doi.org/10.1007/978-3-031-12680-2_14.
Hamilton M, Richardson J. An academic integrity approach to learning and assessment design. J Learn Des. 2007;2(1):37–51. https://eric.ed.gov/?id=EJ903902.
Hanbidge AS, McKenzie A, Scholz KW, Tin T. Academic integrity in the digital era: student skills for success using Mobile technology. In: Yu S, Ally M, Tsinakos A, editors. Emerging technologies and pedagogies in the curriculum. Springer; 2020. p. 335–53. https://doi.org/10.1007/978-981-15-0618-5_20.
Heitman E, Litewka S. International perspectives on plagiarism and considerations for teaching international trainees. Urol Oncol Seminar Original Invest. 2011;29(1):104–8. https://doi.org/10.1016/j.urolonc.2010.09.014.
Howard RM, Jamieson S. The ethics of teaching rhetorical intertextuality. J Acad Ethics. 2021;19(3):385–405. https://doi.org/10.1007/s10805-021-09424-2.
Hunter J, Kier CA. Canadian open digital distance education universities and academic integrity. In: Eaton SE, Christensen Hughes J, editors. Academic integrity in Canada: an enduring and essential challenge. Springer; 2022. p. 249–66. https://doi.org/10.1007/978-3-030-83255-1_13.
Hussein LMT. The effectiveness of teaching educational research course on the development of scientific research skills, academic and personal integrity among female students of Al-Qassim University. Int J Asian Soc Sci. 2017;7(5):Article 5. https://doi.org/10.18488/journal.1.2017.75.392.409.
ICAI. The fundamental values of academic integrity. International Center for Academic Integrity; 2021. https://academicintegrity.org/images/pdfs/20019_ICAI-Fundamental-Values_R12.pdf.
Kenny N, Eaton SE. Academic integrity through a SoTL lens and 4M framework: an institutional self-study. In: Eaton SE, Christensen Hughes J, editors. Academic integrity in Canada: an enduring and essential challenge. Springer International Publishing; 2022. p. 573–92. https://doi.org/10.1007/978-3-030-83255-1_30.
Khan ZR. Academic integrity training module for academic stakeholders: IEPAR framework. J Acad Ethics. 2024;22(1):9–31. https://doi.org/10.1007/s10805-024-09517-8.
Khan ZR, Hemnani P, Raheja S, Joshy J. Raising awareness on contract cheating –lessons learned from running campus-wide campaigns. J Acad Ethics. 2020a;18(2):175–91. https://doi.org/10.1007/s10805-020-09365-2.
Khan ZR, Mumtaz S, Rakhman SS. Mentoring to affect student perceptions of academic integrity. In: Academic language and learning support Services in Higher Education. IGI Global; 2020b. p. 98–118. https://doi.org/10.4018/978-1-7998-2878-5.ch004.
Kier CA, Ives C. Recommendations for a balanced approach to supporting academic integrity: perspectives from a survey of students, faculty, and tutors. Int J Educ Integr. 2022;18(1):22. https://doi.org/10.1007/s40979-022-00116-x.
Kumar V, Verma A, Aggarwal SP. Reviewing academic integrity: assessing the influence of corrective measures on adverse attitudes and plagiaristic behavior. J Acad Ethics. 2023;21(3):497–518. https://doi.org/10.1007/s10805-022-09467-z.
Laka L, Paska PEIN. Academic integrity model development to improve psychological Well-being | Indonesian journal of multidisciplinary science. Indonesian J Multidiscipl Sci. 2023;2(6) https://ijoms.internationaljournallabs.com/index.php/ijoms/article/view/457.
Lancaster T, Clarke R. Dealing with contract cheating: a question of attribution. In: 1st annual Higher Education Academy conference in science, technology, engineering and mathematics; 2012.

Lysiak L. Academic integrity: developing an approach students can own! Pennsylvania Libraries: Res Pract. 2020;8(1):58–63. https://doi.org/10.5195/palrap.2020.217.

MacLeod PD, Eaton SE. The paradox of faculty attitudes toward student violations of academic integrity. J Acad Ethics. 2020;18(4):347–62. https://doi.org/10.1007/s10805-020-09363-4.

Martinson BC, Anderson MS, Crain AL, De Vries R. Scientists' perceptions of organizational justice and self-reported misbehaviors. J Empir Res Hum Res Ethics. 2006;1(1):51–66. https://doi.org/10.1525/jer.2006.1.1.51.

McCabe DL. Cheating among college and university students: a North American perspective. Int J Educ Integr. 2005;1(1) https://doi.org/10.21913/IJEI.v1i1.14.

McCabe D, Pavela G. Some good news about academic integrity. Change Magazine HighLearn. 2000;32(5):32–8. https://doi.org/10.1080/00091380009605738.

McGowan U. Academic integrity: an awareness and development issue for students and staff. J Univ Teach Learn Pract. 2005;2(3):56–66. https://doi.org/10.53761/1.2.3.6.

Morris EJ. Academic integrity: a teaching and learning approach. In: Bretag T, editor. Handbook of academic integrity. Springer; 2016. p. 1037–53. https://doi.org/10.1007/978-981-287-098-8_11.

Morris EJ, Carroll J. Developing a sustainable holistic institutional approach: dealing with realities "on the ground" when implementing an academic integrity policy. In: Bretag T, editor. Handbook of academic integrity. Springer; 2015. p. 1–11. https://doi.org/10.1007/978-981-287-079-7_23-2.

Nikjo P, Vizeshfar F, Zarifsanayee N. Comparison of the effect of training academic honesty using two workshop and virtual training methods on the knowledge and attitude of M. S students. J Educ Health Promo. 2021;10(1):180. https://doi.org/10.4103/jehp.jehp_1145_20.

Ochroch EA, Eckenhoff RG. The role of mentoring in aiding academic integrity. Anesth Analg. 2011;112(3):732. https://doi.org/10.1213/ANE.0b013e318209d4e8.

Okonta P, Rossouw T. Prevalence of scientific misconduct among a Group of Researchers in N igeria. Dev World Bioeth. 2013;13(3):149–57. https://doi.org/10.1111/j.1471-8847.2012.00339.x.

Openo J. Academic integrity and the affordances and limitations of authentic assessment. In: Eaton SE, editor. Second handbook of academic integrity. Springer; 2024. p. 217–31. https://doi.org/10.1007/978-3-031-54144-5_90.

Paine LS. Managing for organizational integrity. Harv Bus Rev. 1994;72(2):106–17.

Pecorari D. Plagiarism, international students, and the second-language writer. In: Bretag T, editor. Handbook of academic integrity. Singapore: Springer; 2016. p. 537–50. https://doi.org/10.1007/978-981-287-098-8_69.

Robinson JA, Glanzer PL. Building a culture of academic integrity: what students perceive and need. Coll Stud J. 2017;51(2):209–21.

Rodrigues M, Silva R, Borges AP, Franco M, Oliveira C. Artificial intelligence: threat or asset to academic integrity? A bibliometric analysis. In: Kybernetes, *ahead-of-print*(ahead-of-print); 2024. https://doi.org/10.1108/K-09-2023-1666.

Solmon MA. Promoting academic integrity in the context of 21st century technology. Kinesiology Rev. 2018;7(4):314–20. https://doi.org/10.1123/kr.2018-0042.

Sotiriadou P, Logan D, Daly A, Guest R. The role of authentic assessment to preserve academic integrity and promote skill development and employability. Stud High Educ. 2020;45(11):2132–48. https://doi.org/10.1080/03075079.2019.1582015.

Spain JW, Robles MM. Academic integrity policy: the journey. Bus Commun Q. 2011;74(2):151–9. https://doi.org/10.1177/1080569911404407.

Steneck NH. Fostering integrity in research: definitions, current knowledge, and future directions. Sci Eng Ethics. 2006;12(1):53–74. https://doi.org/10.1007/s11948-006-0006-y.

Stephens JM. Creating cultures of integrity: a multilevel intervention model for promoting academic honesty. In: Bretag T, editor. Handbook of academic integrity. Springer Singapore; 2016. p. 995–1007. https://doi.org/10.1007/978-981-287-098-8_13.

Stoesz BM, Eaton SE. Academic integrity policies of publicly funded universities in Western Canada. Educ Policy. 2020;0895904820983032 https://doi.org/10.1177/0895904820983032.

Trost K. Psst, have you ever cheated? A study of academic dishonesty in Sweden. Assess Eval High Educ. 2009;34(4):367–76. https://doi.org/10.1080/02602930801956067.

Vasconcelos S, Leta J, Costa L, Pinto A, Sorenson MM. Discussing plagiarism in Latin American science: Brazilian researchers begin to address an ethical issue. EMBO Rep. 2009;10(7):677–82. https://doi.org/10.1038/embor.2009.134.

Wan G, Scott MR. Start them early and right: creating a culture of academic integrity in elementary schools. In: Bretag T, editor. Handbook of academic integrity. Springer; 2016. p. 413–27. https://doi.org/10.1007/978-981-287-098-8_50.

Whitley BE, Keith-Spiegel P. Academic integrity as an institutional issue. Ethics Behav. 2001;11(3):325–42. https://doi.org/10.1207/S15327019EB1103_9.

Role of Policies in Preserving Research and Academic Integrity

9

Ammara Awais, Zaheer Ahmad, and Muhammad Shahid Soroya

9.1 Why Are Policies Important in Preserving Academic Integrity Values?

"Academic integrity policy should be recognized as more than the fences or boundaries that define acceptable behavior. There should be a motivational factor in good policy. What is needed is an inspirational quality that points us to higher expectations and in the case of academic integrity; it can point us to core ethical values that support the community's moral identity." (Wangaard, 2016, p. 434).

Within academic and research settings, well-defined policies are crucial for preserving academic integrity values by clearly outlining ethical standards, norms, and consequences for academic misconduct such as plagiarism, cheating, fabrication, and data falsification (Redman & Caplan 2017). Furthermore, policies create a shared understanding of rules, ensuring that students, faculty, and researchers adhere to honesty, fairness, and accountability principles (Deubel 2018). By enforcing standards and promoting ethical norms and behaviors, policies protect and enforce a culture of academic integrity and uphold the trust placed in educational and research institutes. Scholars have argued that plagiarism will become even more

A. Awais (✉)
Lahore Business School, The University of Lahore, Lahore, Pakistan

Bayes Business School, City, University of London, London, UK
e-mail: ammara.awais@uol.edu.pk

Z. Ahmad
Department of Information Management, The University of Lahore, Lahore, Pakistan
e-mail: zaheer.ahmad@uol.edu.pk

M. S. Soroya
Pakistan Institute of Education, Islamabad, Pakistan

A. K. Abu-Shaheen et al. (eds.), *Introduction to Research Ethics and Academic Integrity*, https://doi.org/10.1007/978-981-96-9678-9_9

possible, especially after the proliferation of artificial intelligence tools and applications, in the absence of well-defined policies (Constantinescu 2024; Eaton 2021). In the section below, we highlight the important reasons for enforcing policies to promote academic integrity within academic and research institutes.

9.2 Upholding Ethical Standards

Ethical standards are a set of principles that guide an individual's behavior and attitude which help ensure fairness, transparency, honesty, accountability, and respect for others in both personal and professional settings. These standards serve as the core foundations for maintaining integrity in research and academic environments (International Center for Academic Integrity 2024). Policies are vital in setting clear expectations for ethical conduct and ensuring that everyone in the organization adheres to principles of respect, fairness, and honesty in recognizing intellectual contributions (Constantinescu 2024).

9.3 Policies as Means of Preventing Plagiarism and Misconduct

Plagiarism undermines the foundational principles of truth, fairness, and integrity in academic environments which can only be prevented by enforcing well-defined policies. Academic integrity policies provide a shield against misconduct (Mulenga & Shilongo 2024). Policy documents provide clear rules and guidelines to prevent misconduct such as cheating, data falsification, and fabrication, and these preventive measures, in the form of policies, become the very foundations of scholarly work.

A culture of integrity promotes habits of proper citations, rigorous research, honest self-reporting, and respecting original intellectual contributions. Furthermore, when faculty and students are well aware of the clearly communicated consequences of misconduct, they will be less likely involved in academic dishonesty and unethical practices (Young 2024). Thus, the policy becomes a fundamental guiding principle and instrumental in shaping and reshaping the culture of accountability and ownership, guiding students and scholars to uphold integrity throughout their academic and research journey (Moriarty 2024).

9.4 Framework for Accountability and Transparency

The policies provide a framework for accountability and transparency within academic institutions, contributing to ethical development and building societal trust in research outcomes (MacLeod 2021). Various scholars agree that well-defined and well-implemented policies ensure that every member of the community is held responsible and accountable for his/her misconduct followed by serious punitive measures (Roik & Kuzmina 2023; Yankova 2024). Their scholarly work promotes the notion that integrity violations are unacceptable and nonnegotiable.

Institutions can clearly explain the expectations, procedures, and punishments associated with academic misconduct by establishing clearly defined and transparent policies (Huberts 2014). Institutional stakeholders are more inclined to respect the integrity of the academic environment when they are aware of how breaches are addressed and the procedures involved in enforcement (Kryvolapchuk & Pluhatar 2024). Transparency also helps to build confidence in the academic community by guaranteeing that everyone follows the same set of norms and is treated equally.

9.5 Ensuring Fairness and Transparency

Fairness and transparency are the key elements in the academic integrity policy framework. Policies are vital in ensuring equal standards for all individuals, fostering integrity and fairness in education (Dodonova et al. 2024). A well-written policy guarantees that all people are held to the same standards of treatment, regardless of experience or position. Impartiality is vital and supports the notion that maintaining integrity is not a privilege but rather a basic requirement for all the stakeholders of the academic and research community.

Institutions can further promote fairness through transparency in the way policies are implemented. Perkins and Roe (2024) highlighted the importance of public accessibility and clarity in policy implementation. Such transparency policy's characteristic not only prevents misconduct but also empowers individuals by providing clear guidelines and assuring fair treatment.

9.6 Consequences of Weak (Ambiguous) or Absent Policies

There is a serious risk to academic integrity when there are unclear and ambiguous policies, particularly in the present era of artificial intelligence. Such unclear policies create a gray area where unethical practices can flourish. Reedy et al. (2021) established that ambiguous policies resulted in diverse interpretations among faculty and students, leading to inconsistent enforcement. Weak policies can also diminish trust and integrity of the academic outcomes (Kryvolapchuk & Pluhatar 2024) which may further damage institutional credibility in the academic community (Dar & Asad 2022).

Students' lack of awareness on academic misconduct can lead to unintentional policy violations. In such an environment, ethical standards are easily compromised leading to the deterioration of the culture of integrity, trust, and respect. It can be challenging for the institution to hold violators accountable for their academic misconduct when policies are not in place (Reedy et al. 2021). Therefore, the institutions must develop unequivocal policies to uphold academic integrity and maintain a positive institutional reputation in academia.

9.7 How Can Policies Impact Academic and Research Communities?

Policies significantly shape academic and research communities by influencing research focus, facilitating interaction between scholars and policymakers, and prioritizing research contents that align with any country's social and economic objectives (Morris 2000; Smith 2010). A culture of collaboration, ethical knowledge exchange practices, innovations, acknowledgment of intellectual contributions, and integrity values can be ensured by establishing comprehensive institutional academic integrity policies. The other important roles of policies in shaping academic and research communities are as follows:

9.8 Policies Cultivate a Culture of Integrity

A well-defined policy not only creates an environment where integrity is celebrated but also presents a balanced approach that includes both punitive and educational elements (Stoesz & Eaton 2022). Policies incorporating ethical principles and cultural contexts foster shared responsibility, shaping the culture of honesty, helping in setting research communities, and cultivating a more acceptable academic culture of integrity (Hughes & Gallant 2016; Yankova 2024). After effective policy enforcement, the resulting culture of integrity becomes the beacon and foundation upon which trust and mutual respect are built within the academic community (Gallant 2008). Wangaard and Stephens (2011) endorsed that policies including core integrity values, commitments from the organization, community engagement and ownership, and curriculum-embedded training can significantly enhance the culture of integrity in the organization.

9.9 Policies Shape Faculty and Student Behavior

When policies are in place, faculty and students can understand the implications of academic dishonesty and are encouraged to model integrity which can positively influence their behaviors and attitudes. The sense of responsibility and ownership can potentially reduce the instances of dishonesty, showcasing faculty as role models, and thereby shaping students' positive behavior and adherence to attitudes towards scholarly conventions.

9.10 Policies Impact Research Quality and Innovation

Policies foster ethical research practices, ensuring originality, proper acknowledgment of the original intellectual contributions, and adherence to ethical principles, significantly enhancing research quality and credibility. Academic integrity policies reduce misconduct and encourage critical thinking, thereby enhancing research quality and encouraging innovative approaches within the Responsible Research and Innovation framework (Das 2019).

9.11 Policies Foster a Culture of Trust and Collaboration

A culture of trust and collaboration can be created and fostered through well-defined policies in place. The resulting trust and collaboration between faculty and students is likely to enhance the reliability and credibility of research outputs which cumulatively improves research quality. It also encourages teamwork, collective problem-solving, and engagement in cooperative learning environments. This culture of effective collaboration, in turn, reinforces trust among each other which can further enhance the credibility of research outputs.

9.12 How to Develop an Effective Policy?

Developing effective policies is vital for the organization to maintain integrity, reputation, and respect while achieving its overall goals. The following are the key aspects to focus on, along with practical steps involved in effective policy development.

9.13 Key Principles for Policy Development

9.13.1 Clarity, Fairness, and Inclusivity

A well-crafted policy is understandable for all stakeholders when it is built on the principles of clarity (easily understandable and unambiguous), fairness (equitable and impartial), and inclusivity (representative of all diverse groups).

9.13.2 Transparency and Accountability

The policy should explicitly mention clear expectations from all relevant stakeholders and the consequences linked with the violations of the policy. The policy should also hold the violators accountable for their misconduct (Young 2024).

9.13.3 Adaptability

The policy must be updated if necessary to changing environments while remaining grounded in the core principles of ethical conduct. It must be culturally sensitive, as seen in the comparison of academic integrity practices in countries like New Zealand

and Bulgaria (Yankova 2024). This becomes necessary as complexity increases, particularly with the proliferation of AI technology and paraphrasing tools (Akhter et al. 2024; Eaton et al. 2023; Mahmud 2024). The Institutional Anti-Plagiarism Committee should continuously take insights from all stakeholders and address various issues, ensuring that academic integrity policies are updated to maintain fairness across diverse educational landscapes. Furthermore, this adaptability principle enables institutions to regularly update policies based on the emerging complex issues and ever-changing educational landscape.

9.13.4 Legal and Ethical Requirements

The policy should mention definitions and details of various forms of plagiarism, lodging complaint mechanisms, investigating plagiarism mechanisms, and guidance for reporting and tracking misconduct cases. It should also include the maximum time duration of the case decision and the kinds of penalties based on the severity of misconduct recommended by the anti-plagiarism standing committee and endorsed by the institutional head. The policy needs to account for the violator's right to appeal and mechanisms to address circumstances where false allegations are put forth to the committee. Consequences of false allegations, and the provision of counseling and support for victims of false allegations, should be mentioned in the policy.

9.13.5 Integrating Technology in Policy Enforcement

Proper guidelines about the ethical use of various technological tools for academic and research activities should be part of the policy to address challenges posed by emerging AI tools and automated paraphrasing tools. Furthermore, a policy can best be implemented if the required tools and anti-plagiarism software have been clearly mentioned in the policy and provided to all stakeholders accordingly.

9.13.6 Stakeholder Involvement in Policy Development

The development of any policy is most effective when it involves input from key stakeholders. Faculty, students, administrators, and even external experts must be part of the policy development process to ensure that policies are wide ranging and address the real challenges faced by the academic community. The word "integrity" comes from the Latin "integrare" (Hoad 1993), which means "to integrate" or "to make whole," as validated and endorsed by Carroll (2016) and Eaton (2021). The 4M Model (Micro, Meso, Macro, and Mega) provides a complete framework and highlights the importance of multi-stakeholder involvement in developing a comprehensive academic integrity policy for the prevention of academic dishonesty in institutions (Gallant 2008; Eaton et al. 2019).

Eaton (2021) revealed that policies are well-positioned to achieve overall objectives when diverse perspectives are included and relevant stakeholders are involved. Promoting a sense of collective ownership and involvement of all stakeholders in the academic integrity policy-making activities can lead to better outcomes in upholding ethical standards, ultimately enabling institutions to enhance their positive branding in academic settings (Mulenga & Shilongo 2024). Multi-stakeholder involvement also helps in anticipating future challenges and potential resistance.

9.13.7 Unambiguous Policy Language

Clear language in the policy document is essential for effective communication and implementation, discussion, and ensuring that every stakeholder understands the expectations and consequences attached to academic misconduct (Perkins & Roe 2024; Yankova 2024). Ambiguous policy language becomes a barrier to understanding, interpreting, and effective policy implementation.

9.14 Mechanisms for Policy Implementation and Monitoring

Implementation is the process of putting policy into action. It is necessary to include a well-defined mechanism for effective policy implementation and monitoring to achieve result-oriented outputs (Roik & Kuzmina 2023; Subramaniam 2024). The policy should outline the required resources, overall objectives to achieve, administrative support, various approaches and models of policy implementation, etc. Similarly, a policy should include a mechanism for monitoring which allows the institutions to improve their implementation strategies based on the feedback and outcomes of the policy.

9.15 Setting Measurement Outcomes

The expected outcomes after the implementation of the integrity policy should be outlined in the policy. The institutions should regularly ask questions such as to what extent accountability, honesty, transparency, and consistent evaluation are being ensured. Additionally, attention must be given to providing the necessary training about the adoption of the policies establishing fairness, transparency, and accountability in practices and measuring the policy outcomes.

9.16 Continuous Review and Adaptation of Policies

Policies should be continuously reviewed and regularly updated to reflect the current educational contexts and technology advancements (Eaton et al. 2023). For example, a recently added feature in Word is "Rewrite" to provide "sentence-level

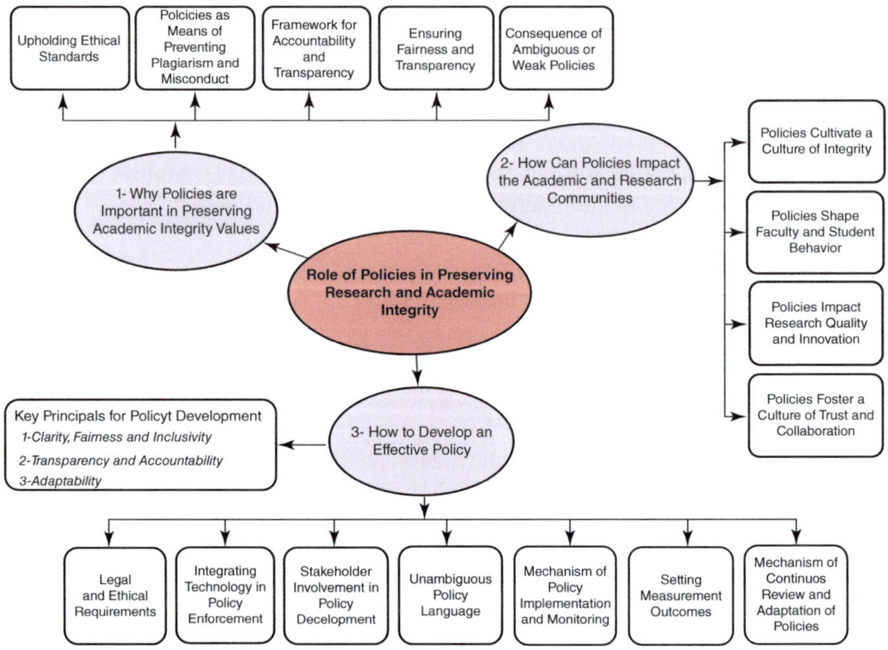

Fig. 9.1 Role of policies in preserving research and academic integrity

suggestions" to help users improve their writing. This feature may challenge our notion of "original work" in ways we can't yet predict (Eaton 2020). This is just one example that supports the idea of continuously reviewing and updating policy to make it relevant to the current time. Redundant and outdated policies negatively influence their enactment and become a hurdle in promoting ethical values among stakeholders. De Maio et al. (2020) indicated the pressing need for regular policy updates, clearly mentioning the frequency of review, and alignment with the evolving academic standards and integrity expectations for effective enforcement and achieving the desired outcomes.

The details of the role of policies in preserving research and academic integrity are exhibited in Fig. 9.1.

References

Akhter N, Suriea U, Soha SA. Understanding the AI integrity in academic excellence: Bangladesh perspective. *Bangladesh*. J Bioeth. 2024;15(2):34–6. https://doi.org/10.62865/bjbio.v15i2.101.

Bertram Gallant T. Academic integrity in the twenty-first century: a teaching and learning imperative. Hoboken, NJ: Wiley; 2008.

Carroll J. Making decisions on management of plagiarism cases where there is a deliberate attempt to cheat. In: Bretag T, editor. Handbook of academic integrity. Singapore: Springer; 2016. p. 199–219.

Christensen Hughes J, Bertram Gallant T. Infusing ethics and ethical decision making into the curriculum. In: Bretag T, editor. Handbook of academic integrity. Singapore: Springer; 2016. p. 1055–73.

Christian Moriarty. So you want an updated (or new!) academic integrity policy. ICAI; 2024 September 23. https://academicintegrity.org/resources/blog/44-2019/december-2019/132-so-you-want-an-updated-or-new-academic-integrity-policy

Constantinescu M. Organisational integrity and responsibility. In: Research handbook on organisational integrity. Edward Elgar Publishing; 2024. p. 394–414.

Dar F, Asad M. Upholding social integrity in higher education institutions. J Educ Educ Dev. 2022;9(2):322–38. https://doi.org/10.22555/joeed.v9i2.743.

Das A. Research integrity in the context of responsible research and innovation (RRI) framework. DESIDOC J Lib Inf Technol. 2019;39(2):82–6. https://doi.org/10.14429/djlit.39.2.13892.

De Maio C, Dixon K, Yeo S. Responding to student plagiarism in Western Australian universities: the disconnect between policy and academic staff. J High Educ Policy Manag. 2020;42(1):102–16. https://doi.org/10.1080/1360080X.2019.1662927.

Deubel P. Punishment or policy change: a case of plagiarism in a dissertation. J Educ Res Pract. 2018;8(1):101–22. https://doi.org/10.5590/JERAP.2018.08.1.08.

Dodonova V, Dodonov R, Voinarovska L, Chornomordenko D, Pavlov Y, Binkivska K, Lobanchuk O. Promoting sustainable education through academic integrity: the habitus and socialization nexus. Eur J Sustain Dev. 2024;13(2):209–9. https://doi.org/10.14207/ejsd.2024.v13n2p209.

Eaton, S. E. (2020). Microsoft's new "Rewrite" feature: a challenge to academic integrity? https://drsaraheaton.wordpress.com/2020/04/20/microsofts-new-rewrite-feature-a-challenge-to-academic-integrity/.

Eaton SE. Plagiarism in higher education: tackling tough topics in academic integrity. Santa Barbara: ABC-CLIO; 2021.

Eaton SE, Chibry N, Toye MA, Rossi S. Interinstitutional perspectives on contract cheating: a qualitative narrative exploration from Canada. Int J Educ Integr. 2019;15(9) https://edintegrity.biomedcentral.com/articles/10.1007/s40979-019-0046-0

Eaton SE, Vogt L, Seeland J, Stoesz BM. Academic integrity policy analysis of Alberta and Manitoba colleges. Can Perspect Acad Integr. 2023;6(1):10.11575/cpai.v6i1.76901.

Hoad TF, editor. The concise Oxford dictionary of English etymology. Oxford: Oxford University Press; 1993.

Huberts L, Six F, van Tankeren M, van Montfort A, Paanakker H. What is done to protect integrity: policies, institutions, and systems. In: The integrity of governance. IIAS series: governance and public management. London: Palgrave Macmillan; 2014. https://doi.org/10.1057/9781137380814_8.

International Center for Academic Integrity. Fundamental values of academic integrity. 2024 September.; https://academicintegrity.org/.

Kryvolapchuk V, Pluhatar T. Academic integrity of scientists: problems of implementation and responsibility in Ukraine. Visegrad J Hum Rights. 2024;2:67–72. https://doi.org/10.61345/1339-7915.2024.2.12.

MacLeod P. A new framework for enhancing (academic) integrity. Can Perspect Acad Integr. 2021;4(2):57–7. https://doi.org/10.11575/cpai.v4i2.74188.

Mahmud S. Fostering academic integrity in the age of artificial intelligence. In: Mahmud S, editor. *Academic integrity in the age of artificial intelligence*. IGI Global; 2024. p. 1–20. https://doi.org/10.4018/979-8-3693-0240-8.ch001.

Morris N. Science policy in action: policy and the researcher. Minerva. 2000;38(4):425–51. https://doi.org/10.1023/A:1004873100189.

Mulenga R, Shilongo H. Academic integrity in higher education: understanding and addressing plagiarism. Acta Pedagogia Asiana. 2024;3(1):30–43. https://doi.org/10.53623/apga.v3i1.337.

Perkins M, Roe J. Decoding academic integrity policies: a corpus linguistics investigation of AI and other technological threats. High Educ Pol. 2024;37(3):633–53. https://doi.org/10.1057/s41307-023-00323-2.

Redman BK, Caplan AL. Improving research misconduct policies: evidence from social psychology could inform better policies to prevent misconduct in research. EMBO Rep. 2017;18(4):511–4. https://doi.org/10.15252/embr.201744110.

Reedy AK, Wurm PA, Janssen A, Lockley A. A community of practice approach to enhancing academic integrity policy translation: a case study. Int J Educ Integr. 2021;17(1):12. https://doi.org/10.1007/s40979-021-00080-y.

Roik O, Kuzmina G. Implementation of the principles of academic integrity in the educational and scientific environment of the Kyiv national university of technology and design. Pedagogy Educ Manag Rev. 2023;4:20–5. https://doi.org/10.36690/2733-2039-2023-4-20-25.

Siva Subramaniam SD. Academic integrity in South Asia: focus on India, Pakistan, and Sri Lanka. In: Eaton SE, editor. Second handbook of academic integrity. Cham: Springer; 2024. https://doi.org/10.1007/978-3-031-54144-5_88.

Smith K. Research, policy and funding–academic treadmills and the squeeze on intellectual spaces 1. Br J Sociol. 2010;61(1):176–95. https://doi.org/10.1111/j.1468-4446.2009.01307.x.

Stoesz BM, Eaton SE. Academic integrity policies of publicly funded universities in western Canada. Educ Policy. 2022;36(6):1529–48. https://doi.org/10.1177/0895904820983032.

Wangaard DB. Practices to support developing academic integrity in secondary school students. In: Bretag T, editor. Handbook of academic integrity. Singapore: Springer; 2015. https://doi.org/10.1007/978-981-287-079-7_34-1.

Wangaard DB, Stephens JM. Creating a culture of academic integrity: a toolkit for secondary schools. Minneapolis: Search Institute Press; 2011.

Yankova D. Academic integrity policies in HEIs across cultures. English Stud NBU. 2024;10(1):5–20. https://doi.org/10.33919/esnbu.24.1.1.

Laura Young. How can leaders in higher education build an academic integrity policy? Turnitin. 2024 September 23.; https://www.turnitin.com/blog/how-can-leaders-in-higher-education-build-an-academic-integrity-policy.

10 Developing a Culture of Integrity Among Students and Academics

Sivapriya Ramakrishnan and Shivadas Sivasubramaniam

10.1 Overview

This chapter focuses on the ways to develop and improve the culture of academic and research integrity among students and academics at various levels of education. It explores different approaches to incorporate integrity education as a cocurricular activity from secondary education, higher education, and beyond by continuous scaffolding to enhance individuals' own integrity as a part of their attributes and apply it throughout their career and lifestyle.

10.2 Academic Integrity and Its Sustainability in Higher Education

Academic integrity serves as a fundamental pillar in higher education, encompassing virtues such as honesty, ethical conduct, and responsible scholarship. Upholding values like honesty and intellectual authenticity is crucial for the sustained credibility and effectiveness of academic institutions. The collective commitment of students, faculty, administrators, and stakeholders is essential to foster a culture that values academic integrity. This shared commitment ensures that higher education remains a bastion of quality education and contributes to societal progress over the long term.

S. Ramakrishnan (✉)
Department of Physiotherapy, College of Health Sciences, University of Sharjah, Sharjah, United Arab Emirates
e-mail: sramakrishnan@sharjah.ac.ae

S. Sivasubramaniam
Deputy Dean, School of Life and Health Sciences, University of Roehampton, London, UK
e-mail: S.Sivasubramaniam@roehampton.ac.uk

A. K. Abu-Shaheen et al. (eds.), *Introduction to Research Ethics and Academic Integrity*, https://doi.org/10.1007/978-981-96-9678-9_10

10.3 Raising Awareness and Providing Education Within the Academic Community

Initiatives such as workshops, seminars, and orientation sessions can enlighten students, faculty, and staff about the significance of academic integrity, emphasizing ethical behavior, proper citation, and responsible research practices. Clear policies and expectations form the second set of strategies, establishing guidelines for students and faculty regarding academic misconduct, plagiarism, and cheating. Institutional commitment, highlighted in the third strategy, emphasizes the importance of integrating academic integrity as a core value in the mission statement and policies of the institution, fostering a top-down approach to ethical conduct.

10.4 Specific Measures for Promoting Academic Integrity Within the Academic Environment

Faculty members, as role models, are encouraged to incorporate discussions on academic integrity into their courses and research projects. Diverse assessment methods, including personalized assignments and practical exams, aim to discourage cheating and promote original thought. Technology integration, such as plagiarism detection software and secure online exam platforms, is also advocated, along with educating students on the ethical use of technology. Support services, including writing centers and academic counseling, are crucial for helping students navigate the demands of higher education in an honest and informed manner. Peer accountability, continuous evaluation, and the celebration of ethical behavior further contribute to a robust framework for academic integrity.

10.5 Global Collaboration in Promoting Academic Integrity

Engaging in international efforts to share best practices, resources, and research with other institutions establishes common standards for upholding integrity globally. Ethical leadership programs, the twelfth strategy, aim to equip students with the skills needed to make ethical decisions not only in their academic pursuits but also in their professional lives. In summary, these comprehensive strategies encompass a multifaceted approach, ensuring that academic integrity is not only preserved but also strengthened as a guiding principle in higher education.

10.6 Historical Perspectives of Academic Integrity

Historically speaking, academic integrity is a phenomenon that is expected from scholars rather than enforced. In other words, it was linked to a person's moral code focusing on honor, duty, pride, loyalty, and above all self-esteem with honesty

(Gallant 2008). This formed a part of developing a character or behavior, which was expected to be self-owned by professionals. Any incidences of dishonesty were usually expected to be self-reported. This might have resulted in fewer checks to maintain academic integrity. Therefore, integrity checks varied across institutions and largely relied on traditional methods of detecting plagiarism using academic wisdom. These methods include manual reading and comparison of student papers, the use of academic honor codes and policies, and occasional suspicion-based investigations (some of which are elaborated below):

(a) **Manual Detection**: Institutions and faculty use manual detection methods to utilize their knowledge and experience to detect plagiarism by comparing students' work. This method is not foolproof; it is time-consuming owing to the increase in the number of students and the pressure given to the faculty to focus more on research, and the time spent on teaching/learning is less.
(b) **Suspicion-Based Investigation**: In cases where plagiarism is suspected or reported, faculty members or administrators would investigate further by comparing suspicious sections of a student's work with potential sources. This process was subjective and relied heavily on the intuition, student behavior, and experience of the person conducting the investigation and prejudice.
(c) **Academic Honor Codes:** Educational institutions had been with honor codes that emphasized academic integrity and reduced the number of academic misconducts. Students were expected to follow these honor codes and were often required to undertake the induction courses on academic integrity and sign an honor pledge at the beginning of each semester, assignment, or exam. An honor code in higher education is a set of principles and rules that students and faculty members agree to abide by to promote academic integrity, and ethical behavior, and develop a sense of community within the educational institution. The primary goal of an honor code is to establish a culture of trust, respect, and responsibility among all members. Honor codes typically cover various aspects of academic life, including exams, assignments, research, and interactions among students and faculty. Let's see this in detail highlighting some key components and benefits of an honor code.

10.7 Components of an Honor Code

Pledge or Statement Students and faculty often make a formal pledge to uphold the honor code's principles. This pledge may be recited in a ceremonial setting (or during important events) and sometimes written at the beginning of assignments and exams.

Academic Integrity The honor code outlines what constitutes academic integrity, such as not engaging in cheating, plagiarism, or any form of academic dishonesty. It also encourages proper citation and original work.

Collaboration Guidelines The code may provide guidance on acceptable levels of collaboration on assignments, projects, and exams, clarifying when group work is permitted and when independent effort is expected.

Reporting Procedures Procedures for reporting violations and addressing breaches of the honor code are usually outlined. This may involve a student reporting to another student or a faculty member addressing concerns.

Consequences The honor code should clearly state the consequences of violating its principles. These consequences can range from warnings and educational interventions to more severe penalties like academic probation or expulsion.

Benefits of Honor Code in Developing a Culture of Integrity An honor code fosters a sense of responsibility and honesty among students providing the opportunity for them to own and promote a culture where academic achievement is based on genuine effort and merit. A strong honor code builds trust between the students and the faculty, as everyone knows the expectations and principles that guide their academic interactions. By emphasizing ethical conduct, an honor code prepares students for ethical decision-making in their professional careers and personal lives. The shared commitment to the honor code creates a sense of community among students, reinforcing positive behavior and mutual respect. Studies have suggested that institutions with well-implemented honor codes experience lower rates of cheating and academic dishonesty (Raman and Ramlogan 2020; Zhao et al. 2023). Upholding the honor code prepares students to face challenges and dilemmas in their careers and personal lives with a strong ethical foundation. Institutions with a reputation for academic integrity and a strong honor code can attract students, faculty, and resources more effectively.

Policies to Maintain Academic Integrity Implementing and maintaining an honor code requires ongoing effort, communication, and education. Faculty, staff, and students need to work together to ensure that the honor code remains relevant, understood, and respected by all members of the academic community. Regular discussions, awareness campaigns, and orientations can help reinforce the importance of the honor code in maintaining integrity within higher-education institutions. Maintaining academic integrity is essential for upholding the value and credibility of education. Educational institutions often implement various policies and strategies to promote and enforce academic integrity. Some of the common policies and approaches that institutions are summarized in Table 10.1.

Table 10.1 Example summary collections of institutional policies, procedures, and approaches

Institutional undertakings	Faculty-based approaches	Expectations from the students
Honor code: An honor code sets clear expectations for ethical behavior and academic integrity for the institution. It establishes a shared commitment among students and faculty to uphold these principles.	**Text-matching tools:** Widely used to scan assignments, papers, and research for instances of potential plagiarism or improper citation. Examples of such tools include Turnitin, Grammarly, and Copyscape.	**Education and awareness programs**: Regular workshops, seminars, and orientations to educate students about the importance of academic integrity, proper citation practices, and the consequences of dishonesty
Leadership and role modeling: Institutional leaders, including administrators and faculty, must exemplify integrity in their actions and decisions. Their behavior sets the tone for the entire institution.	**Showcasing good practices**: Integrity-related articles, posters, and other outputs from the faculty members are displayed. Continuous training opportunities such as integrity seminars and workshops for students and staff	**Providing student opportunities**: Students are empowered to take leaderships in organizing events (such as international day of action against plagiarism/contract cheating)
Clear academic integrity guidelines: Institutions providing detailed guidelines on what constitutes plagiarism, cheating, and other forms of academic dishonesty. These guidelines help students understand expectations and consequences. **Transparent policies:** Clearly communicate academic integrity policies, procedures, and consequences in a way that is easily accessible to all members of the community.	**Assignment variety for active learning:** Creating a variety of assignments (e.g., essays, projects, presentations, exams) can make cheating and plagiarism more challenging, as each assignment requires unique approaches. Assignments that expect knowledge/skills application as opposed to "knowledge reproduction"	**Open communication**: Fostering an environment where students feel comfortable discussing their concerns or confusion about academic integrity. Clear communication and flexibility to prevent unintentional violations.
Citation styles: Institutions mandating specific citation styles (such as APA, MLA, Chicago, etc.) to ensure proper attribution of sources. Students are expected to follow these styles in their assignments.	**Minimizing the chances of misconducts:** Limited time windows for online assignments and exams, reducing the opportunities for students to collaborate or search for answers.	**Personalized assessments**: Designing assessments that require critical thinking and personal reflection makes it more difficult for students to find ready-made answers online. **Feedback and reflection:** Provide constructive feedback on assignments, emphasizing the value of improvement and growth over grades alone. Encourage students to reflect on their learning process.

(continued)

Table 10.1 (continued)

Institutional undertakings	Faculty-based approaches	Expectations from the students
Proctored exams: For online courses, proctored exams use software or human proctors to monitor students during exams, reducing the likelihood of cheating.	**Interactive learning**: Engage students in active learning experiences that encourage critical thinking, problem-solving, and personal reflection. Such experiences reduce the temptation to resort to dishonest practices.	**Faculty engagement**: Encourage faculty members to discuss academic integrity in class and integrate discussions about ethics into the curriculum.
Inclusive community: Promote inclusivity and diversity in the academic community. When students feel respected and valued, they're more likely to uphold ethical standards.	**Faculty engagement**: Faculty members play a crucial role in setting up platforms for equality, diversity, and inclusivity. Celebration of honesty: Recognize and celebrate students who exemplify integrity in their academic work and behavior.	**Students' initiatives:** Student union and individual subject-specific journal clubs and social activities to promote integrity
Institutional support: Offer resources such as writing centers, research assistance, and tutoring to help students succeed without resorting to dishonesty.	**Enforcement:** Ascertain the institutional support measures are incorporated in subject-specific contexts. **Continuous improvement**: Regularly assess the effectiveness of integrity initiatives and seek feedback from students and faculty on how to improve the culture of integrity.	**Empowerment:** Empower students to take ownership of their learning and to pursue academic excellence with integrity.

10.8 Assessment Methods Used in Maintaining Academic Integrity

Maintaining academic integrity in assessments is crucial to ensure that students' learning outcomes accurately reflect their knowledge and skills. Here are some examples of different assessment methods that educational institutions use to promote and maintain academic integrity (Table 10.2). Institutions often use a

Table 10.2 Examples of different assessment methods to enhance academic integrity

Traditional in-person exams	**Proctored exams:** Exams are monitored by invigilators, either in person or remotely through video proctoring, to deter cheating. **Seating arrangements**: Randomized seating in exam halls minimizes opportunities for collaboration. Unique exam versions: Multiple versions of the same exam with scrambled question orders or different questions reduce cheating.
Online assessments	**Timed exams:** Setting time limits on online exams reduces the possibility of looking up answers. **Randomized questions**: Shuffling the order of questions and answer choices for each student reduces the effectiveness of sharing answers. **Question pools**: Drawing questions from a larger pool ensures that students receive different sets of questions. **Lockdown browsers:** Specialized browsers lock down the testing environment, preventing access to external resources. **Remote proctoring:** Using webcam and microphone monitoring software to observe students during online exams
Authentic assessments	**Projects**: Assignments that require critical thinking, application of knowledge, and personal interpretation discourage plagiarism. **Case studies:** Presenting real-world scenarios for analysis and problem-solving tests students' ability to apply concepts. **Portfolios:** Collecting a range of student work continuously over time showcases growth and learning outcomes via reflective learning.
Performance-based assessments	**Oral exams (viva voce)**: Students explain their understanding verbally, reducing the ability to use external resources. **Practical/spot tests:** Hands-on tests in fields such as science, engineering, and art demonstrate skills and knowledge.
Group assessments	**Peer evaluation:** Incorporating peer evaluations encourages students to assess their group members' contributions. **Individual components:** Including individual components in group projects helps ensure each student's accountability.
Open-book assessments	**Complex questions:** Designing questions that require critical thinking and analysis rather than simple fact recall encourages genuine engagement.
Reflection and journals	**Reflective essays**: Requiring students to reflect on their learning experiences or personal growth encourages original thought. **Learning journals or diaries:** Regular journal entries document students' learning journey and insights.
Personalized assessments	**Personal experiences**: Assignments that draw on students' personal experiences make cheating more difficult. **Opinion-based assignments**: Assignments that ask for personal opinions or interpretations discourage copying.
Verbal presentations	**Individual presentations:** Requiring students to present their findings or perspectives orally showcases their understanding. **Fact-check questioning:** Follow-up questions from the audience or instructor can probe deeper into a student's knowledge.

combination of these methods to create a comprehensive assessment strategy that aligns with their goals and educational approach. It's important to regularly review and adapt these methods to maintain their effectiveness and address new challenges in preserving academic integrity.

10.9 The Role of Automated Feedback to Enhance Academic Integrity

Automated grading and feedback systems have emerged as potent tools with the potential to significantly bolster academic integrity in higher education. The positive impacts of these systems are noteworthy. First and foremost, they offer consistency in grading, mitigating the risk of biased assessments and fostering fairness. Additionally, their ability to provide timely and objective evaluations ensures that students receive prompt feedback based on predetermined criteria, minimizing subjective biases that could compromise the integrity of the assessment process. The anonymity maintained by automated grading during evaluation further contributes to impartiality, creating an environment conducive to ethical academic practices. Moreover, the efficiency gained allows instructors to allocate more time to personalized feedback and addressing concerns related to academic integrity.

Addressing academic integrity concerns head-on, automated systems play a pivotal role in detecting plagiarism, a pervasive issue in educational settings. By comparing submitted work against a comprehensive database, these systems can identify instances of potential plagiarism and encourage proper citation practices. Furthermore, automated feedback systems extend their impact by providing citation guidance, flagging sections that appear unoriginal, and offering educational insights to students about the importance of original work. In addition, they can detect patterns of collaboration that may indicate unauthorized sharing of answers, supporting a proactive approach to maintaining academic honesty. Generating detailed reports on integrity issues, these systems empower instructors to take appropriate actions and reinforce ethical standards.

However, it is essential to acknowledge the considerations and limitations associated with automated grading and feedback systems. While their consistency and objectivity are strengths, these systems might lack a nuanced contextual understanding of a student's work, leading to potential misinterpretations. Subjective evaluations, particularly in assignments requiring interpretative analysis, may pose challenges for automated systems. Additionally, the feedback provided may lack the depth and personalized touch inherent in human assessments, potentially hindering the development of critical thinking and writing skills. The risk of false positives in plagiarism detection algorithms and the adaptability of systems to newer forms of academic misconduct further highlight the need for caution.

In conclusion, the incorporation of automated grading and feedback systems offers a promising avenue to reinforce academic integrity. However, a balanced approach is crucial, emphasizing their role as tools to enhance the learning experience rather than replacing the essential elements of human guidance, contextual understanding, and nuanced assessment. These systems, when used judiciously alongside instructor vigilance and clear communication about ethical behavior, can contribute significantly to the cultivation of a culture of integrity within educational institutions.

10.10 The Future of Integrity in Education and Beyond

The future of integrity in education is intricately linked to technological progress, societal shifts, and the dynamic nature of learning. With advanced AI and machine learning driving the evolution of plagiarism detection tools and cheating prevention measures, technology is set to play a pivotal role. The surge in online and hybrid learning models necessitates innovative approaches to uphold academic integrity in virtual environments. Personalized learning demands adaptable assessment methods, while blockchain technology ensures secure credential verification. The integration of ethics education becomes paramount as AI influences various industries, including education. Global collaboration and the sharing of best practices are expected to grow in promoting academic integrity. Traditional exam formats may give way to open-book assessments, emphasizing real-world application and critical thinking. A focus on soft skills aligns education with the evolving job market, emphasizing ethical decision-making. Teaching digital literacy equips students to navigate online resources responsibly. Transparent AI tools for grading and feedback enhance student understanding. Educational institutions will continue to shape societal values and ethical behavior. Integrity in lifelong learning and professional development becomes increasingly vital. Adaptive ethics education programs will address emerging challenges. Human-centered approaches, alongside technological innovation, remain crucial for fostering integrity. Education will extend its scope to address broader ethical considerations, preparing students for ethical engagement in a complex world. Therefore, the future requires a comprehensive approach, combining technology, pedagogy, and a commitment to ethical values for the enduring foundation of integrity in education.

References

Gallant T. "Revisiting the past: the historical context of academic integrity". Academic integrity in the twenty-first century. ASHE High Educ Rep. 2008;33(5):13–31. ISBN 9780470373668.

Raman V, Ramlogan S. Academic integrity and the implementation of the honour code in the clinical training of undergraduate dental students. Int J Educ Integr. 2020;16:9. https://doi.org/10.1007/s40979-020-00058-2.

Zhao L, Peng J, Yang X, Yan W, Ke S, Dong LD, Li Y, Ma J, Lee K. Effects of honor code reminders on university students' cheating in unproctored exams: a double-blind randomized controlled field study. Contemp Educ Psychol. 2023;75:ISSN 0361-476X. https://doi.org/10.1016/j.cedpsych.2023.102213.

11 Case Studies

Leen Abo Safieh, Lubna Ansari Baig,
and Sivapriya Ramakrishnan

11.1 First Case: Overlooked Ethical Breaches

11.1.1 Case Background

A reputed university research team conducted a study on a new therapeutic intervention for poststroke rehabilitation, focusing on improving motor function and quality of life of stroke patients. The study involved physical therapists and patient participants in biweekly therapy sessions over 6 months. Despite Institutional Review Board (IRB) approval, several overlooked ethical issues emerged during this research.

1. While participants initially signed a consent form, many were unaware of the study's extensive time and physical requirements. Some participants continued despite discomfort, feeling they had no choice.
2. Recruitment relied on participant referrals through existing physiotherapy clinics. Several participants felt obligated to join, fearing it might affect their therapeutic relationship or lose priority in future appointments.

L. A. Safieh (✉)
Research Center, Riyadh Second Health Cluster, Riyadh, Kingdom of Saudi Arabia
e-mail: lsafieh@kfmc.med.sa

L. A. Baig
Department of Medical Education, University College of Medicine and Dentistry, University of Lahore, Lahore, Pakistan

S. Ramakrishnan
Department of Physiotherapy, College of Health Sciences, University of Sharjah, Sharjah, United Arab Emirates

A. K. Abu-Shaheen et al. (eds.), *Introduction to Research Ethics and Academic Integrity*, https://doi.org/10.1007/978-981-96-9678-9_11

3. Data, including video recordings of therapy sessions, were stored on shared drives accessible to both research and clinical staff.
4. Throughout the study, participants received no updates on the study's preliminary findings or progress.
5. The study's design emphasized physical outcomes but did not monitor the participants' emotional well-being, which is especially relevant in poststroke populations facing rehabilitation challenges.

Question: What are the possible integrity principles compromised in this research?

Answer: Integrity Principles And Ethical Analysis.
In physical therapy research, upholding research integrity relies on several key ethical principles. Respect for persons emphasizes that informed consent is an ongoing, active process that protects participants' autonomy and their right to withdraw at any point, particularly if discomfort arises. This principle ensures that participants receive full, accurate information to make informed decisions throughout the study.

The principle of justice and fairness calls for unbiased recruitment strategies that avoid real or perceived coercion; using independent recruitment channels, rather than clinical relationships, promotes impartiality and equitable participation. Responsibility in data management is essential for safeguarding participant privacy, requiring encrypted, access-restricted data storage and strict adherence to GDPR standards, which reinforces trust and confidentiality. Furthermore, transparency with participants involves regular debriefing sessions that keep participants informed and engaged, empowering them to understand their contributions and the study's impact on their care. Lastly, beneficence in participant well-being requires researchers to actively minimize harm and enhance participants' physical and emotional well-being, fostering a supportive environment where participants feel respected and valued.

11.1.1.1 Guidelines to Handle This Situation

1. **Ongoing Consent Checks:** To uphold participant autonomy, researchers should regularly confirm participants' willingness to continue, particularly when discomfort arises. Brief, periodic check-ins reinforce informed consent as a continuous process, supporting autonomy and ethical engagement.
2. **Fair Recruitment Practices:** Recruitment should avoid coercion by employing neutral, independent channels, thus preventing any perceived obligation. By not utilizing clinical relationships, researchers ensure justice, allowing participants to make impartial decisions.
3. **Secure Data Management:** Responsibility in data handling mandates encrypted storage and access restricted solely to authorized personnel. Regular audits and strict guidelines of data protection policy compliance are to be maintained to protect confidentiality and enhance participant trust.
4. **Transparency Through Debriefing:** Scheduled debriefing sessions provide participants and their families with ongoing insights into study progress and

preliminary results. This transparency respects participants as informed collaborators and supports ethical integrity.

5. **Well-Being Assessments:** Beneficence requires the regular monitoring of participants' physical and emotional health. If the research is overwhelming to the participants, they have to be reminded about the right to withdraw from the study. Through systematic assessments or check-ins, researchers can promptly address any issues, safeguarding participant welfare throughout the study.

11.2 Second Case: Plagiarism

11.2.1 Case Background

In a leading university, two professors from pharmaceutical sciences were charged with plagiarism. The head of the department and another professor in the same department published a paper "Grapefruit juice—drug interactions" in a local journal indexed in MEDLINE/PubMed in 2005. The two professors supervised a junior ad hoc teacher who was also included in the authorship of the paper. The relevant authorities (Higher Education Commission) received an email from a professor in the United States that the same paper had been published in 1998 in the "British Journal of Clinical Pharmacy" authored jointly by three professors from the United States. The university fired the junior ad hoc teacher and set up a committee to investigate the case thoroughly and the professors were suspended till the finalization of the committee report.

The university at that time did not have a clear-cut policy on contravention of academic integrity. The university sought the opinion from the higher courts of the country; the case was argued in court and a case of plagiarism was proved against the two professors.

Question: What should be the action taken by the university?

Answer The judges ruled that since the university had no law on plagiarism, also that plagiarism has not been defined, misconduct has been defined under section 2(2) of the University Employees (Efficiency and Discipline) Ordinance 1962.

The university demoted the professors and fired the junior teacher.

11.3 Third Case: Excessive Self-Citation

11.3.1 Case Background

Professor XY, a distinguished faculty member at a university made headlines after being accused of engaging in hyper-citation (excessive self-citation or over-citation of a particular set of works, often aimed at artificially inflating the perceived impact of a scholar's research output practices). This allegation emerged after a study analyzing citation patterns in scientific literature revealed an unusual citation frequency of his own published works in recent articles authored by him and his collaborators. His research, which was mostly on biomedical sciences, had gained significant recognition in the scientific community. However, the increasing number of citations to his work, often without substantial new contributions or direct relevance to the cited papers, raised concerns about potential self-promotion and unethical citation practices.

The issue was brought to light in a prominent journal, which discussed the ethical implications of self-citation and hyper-citation in academic publishing. The editorial highlighted the growing concern within the scientific community about researchers manipulating citation metrics to enhance their academic reputations and funding opportunities. The journal raised the ethical dilemma of whether hyper-citation practices, while not necessarily violating formal publishing guidelines, undermine the integrity of the peer review system and distort the true value of scientific contributions. The editorial argued that excessive self-citation could potentially harm the credibility of the research community and obscure the work of other scientists deserving of recognition.

Question: What are the ethical implications of excessive self-citation?

Answer In this case, the ethical implications include manipulation of metrics, misleading academic records, impact on emerging scholars, transparency, and accountability. The case raised questions about the transparency of citation practices. It highlighted the need for clear ethical guidelines regarding self-citation and the responsibilities of researchers and publishers to ensure that citation practices are used appropriately and in ways that foster academic integrity.

The controversy surrounding Professor XY serves as a reminder of the ethical responsibilities of researchers in maintaining the integrity of the scientific publishing process. While self-citation is a normal part of academic life, hyper-citation can undermine the credibility of both the individual researcher and the broader scientific community. As the journal editorial board concluded, journals, institutions, and funding bodies must work together to develop clearer standards for citation practices and promote ethical behavior in the academic world.

11.4 Fourth Case: Use of Fabricated Data in Research

11.4.1 Case Background

A postdoctoral fellow, Dr. X noticed the use of fabricated data in multiple publications by one of the senior researchers within his group. He did this to show positive results in favor of a new drug that was still under development by a pharmaceutical company, the same company that funds their research.

Question: How to tackle research misconduct?

Answer

1. *Identify the issue*: Dr. X detected inconsistencies in the datasets and images across multiple publications attributed to senior researchers. Other colleagues within the team expressed skepticism, but no one had concrete evidence.
2. *Collect evidence*: Dr. X gathered as many relevant materials as he can, such as lab notes, journals, raw data files, and earlier drafts that point to irregularities or dubious procedures. Thorough documentation is crucial to maintaining the security and confidentiality of the evidence and pinpointing the discrepancy in the data reported.
3. *Consultation*: Seeking consultation is important to assure fairness and professionalism in handling the situation. It is important to identify a trusted mentor from outside the team to provide advice. The mentor must review all evidence collected and guide how to proceed responsibly without jeopardizing Dr. X's career.
4. *Internal reporting*: Dr. X chose to inform the organization's Research Integrity Officer (RIO) about his observation. To guarantee that the problem is formally recognized and resolved, this is accomplished through the proper reporting channel: intranet links, electronic forms, etc. Dr. X presented the gathered data and raises questions regarding patient safety and study integrity.
5. *Internal investigation*: A formal investigation was initiated at this point by initiating an inquiry committee. A committee comprising individuals with relevant expertise and impartiality was formed. The committee examined all the evidence submitted, speaks with the senior researcher and Dr. X, and determined whether the suspects' statements are credible.
6. *Protection for whistleblower*: Under institutional policies, Dr. X is guaranteed protection as a whistleblower during the entire procedure. To avoid reprisals, confidentiality is upheld, and Dr. X received guidance on how to handle any difficulties that may arise throughout the investigation.
7. *Outcome and consequences*: If misconduct is evident, then appropriate actions are taken against the senior researcher which may include retraction of fraudulent publications, disciplinary measures, or even legal consequences. As a standard protocol, the organization reinforces and trains researchers in ethical procedures and upholds policies about research integrity.
8. Dr. X was encouraged to continue adopting a culture of integrity within the lab, advocating for transparency in research, and participating in workshops on ethical research practices. Regular audits and training programs must be implemented to mitigate future risks of misconduct.

11.5 Fifth Case: Ghostwriting in Clinical Research Practice

11.5.1 Case Background

Ghostwriting in the context of medical research describes the practice of either not recognizing the real contributors or giving credit to others who did not make a substantial contribution to the literature or research. Concerns about research ethics and integrity in the medical industry are seriously raised by this approach. The integrity of medical research relies on transparency. Ghostwriting obscures the contribution of researchers, making it difficult to assess the credibility of the findings and the motivations behind publishing certain results, especially in relation to commercial interests. In one case, a scientist accepts an offer from a pharmaceutical company to write and publish data related to one of their products that he is also still doing his research on. They promised him that this article would have a huge impact on the scientific world and that his name as a first author would be cited so many times. Being busy with his research, he accepted the offer, and the data would be published. He then faces so many issues regarding the reliability and consistency of the data presented and the conclusions made, and he finds himself in a situation that he cannot explain.

Question: What are the key points to be considered related to ghostwriting and medical research integrity?

Answer

- Genuine authorship should reflect substantial contributions to the conception, design, execution, or interpretation of the research. Ghostwriting undermines this principle by misrepresenting who is responsible for the research and can lead to a lack of accountability.
- Some pharmaceutical companies engage professional writers to produce articles that promote their products, which can lead to biased results and an increased risk of misinformation in the literature. Such practices can distort clinical practice and patient care.
- Ghostwritten articles may contribute to clinical practice guidelines, potentially leading to recommendations based on misleading information. This can affect patient treatment strategies and undermine public trust in the medical literature.
- Increasing scrutiny from regulatory bodies, medical journals, and academic institutions has led to the implementation of stricter guidelines regarding authorship and disclosure of conflicts of interest.
- A significant consequences can result due to ghostwriting can lead to significant consequences, including retraction of published articles, damage to reputations, loss of funding, and legal ramifications, particularly if the work is found to be fraudulent or misleading.
- Adhere to the institutional and respectful journal rules and regulatory measures such as peer review processes, checks for authorship legitimacy, and disclosures of financial interests.

11.6 Sixth Case: Publishing in Predatory Journals

11.6.1 Case Background

A junior researcher was given the task by his team to handle submitting an article for publication. The researcher started searching for related journals that would accept his article. Following the excessive search, the researcher found a journal with an impact factor that publishes relatively quickly and for small fees in comparison with other journals. Being keen to impress his team in handling this task quickly, he decided to go ahead and submit the article for publication. The researcher was happy when he heard back from the journal within the first week and was even happier when there were no scientific comments or edits suggested. He was asked to pay fees and the article was published in less than 2 months. It was only when one of the authors needed to use this published article for promotion application that they discovered that this journal is predatory and would not be considered. The unfortunate incident is that they cannot withdraw it nor they can publish it in another place.

Question: What are the main things to look for to avoid publishing in predatory journals?

Answer

- History of the journal and presence on respectful scientific databases.
- Check the journal editors and the review process.
- No one can write a perfect article; if no comments are given by the reviewer on your article, this should make you worry, not happy.
- Ignore emails with invitations to submit your article; the majority of these are predatory.
- Consult your senior collogues and experts before submitting your article.
- Encouraging open research practices and the sharing of research data can also help enhance and educate junior researchers regarding the publication process.

11.7 Seventh Case: Misuse of AI in Research

11.7.1 Case Background

A team consisting of diverse areas of research interests decides to jointly write a research proposal and submit it for funding. The task was distributed according to each member's areas of expertise. The PI screened through the proposal and applied for funding approval.

Following the reviewers' process, the application was rejected due to the 100% use of AI in certain parts of the proposal. The principal investigator was not aware that one of the co-investigators did this for the part assigned to him.

Question: What are the points to consider when using AI to uphold research integrity?

- Always declare the use of AI in any setup, proposal, publication, etc.
- Refer to the application and institutional guidelines regarding the use of IA.
- The principal investigator must thoroughly review the proposal.
- The team must understand the consequences of rejecting a proposal and how it reflects on the entire team. The use of AI should augment human expertise rather than replace the effort involved in crafting substantial parts of a research proposal.

Glossary

4M framework A model proposed by Kenny and Eaton (2022) for categorizing strategies at four levels (micro, meso, macro, mega) to address academic integrity issues comprehensively.

Academic honesty and accountability The ethical commitment to truthfulness, integrity, and responsibility in all aspects of learning, teaching, and research.

Academic integrity A commitment to honesty, trust, fairness, respect, responsibility, and courage in academic work, essential for credible and valid research.

Academic misconduct Violations of academic integrity, such as plagiarism, data fabrication, falsification, ghost authorship, and ethics breaches, which undermine academic and scientific credibility.

Academic transparency The open and honest sharing of information, methods, and findings in academic work to ensure accountability, reproducibility, and trustworthiness in learning and research.

Academic writing development The process by which students cultivate their writing style, habits, and skills, ideally grounded in principles of academic integrity from early education stages.

Anonymity A process in research where participants' identities are not known to the researcher or anyone else, ensuring the highest level of privacy protection.

Artificial intelligence Using technology to automate tasks that normally require human intelligence.

Authentic assessment An assessment approach designed to promote academic integrity by creating tasks that mimic real-world scenarios and encourage genuine student engagement.

Automated feedback Feedback provided by an automated system or tool, often used in online learning platforms. It helps students assess their performance or understanding of material based on preprogrammed criteria, offering instant responses to assignments, quizzes, or exams.

Beneficence The ethical principle that requires researchers to act for the benefit of participants, ensuring the research has positive impacts and minimizing potential harm.

A. K. Abu-Shaheen et al. (eds.), *Introduction to Research Ethics and Academic Integrity*, https://doi.org/10.1007/978-981-96-9678-9

Cheating/fraud Cheating is generally referred to as copying in exam, copying from another student's assignment, etc. Fraud is specifically related to financial embezzlement.

Cocurricular Activities, programs, and learning experiences that complement the formal academic curriculum. These activities contribute to the overall educational development of students, such as sports, arts, clubs, and leadership programs.

Collaboration Is a group of two person or more working together toward achieving a research goal.

Collusion Collaboration between individuals that results in unauthorized aid or support, commonly observed as a form of academic misconduct among students.

Compilation plagiarism When a work of research is filled with well-cited quotations and references but fails to contribute any original thought or idea.

Complete plagiarism Copying someone else's entire work and pretending to be its original author. This is the most blatant form of plagiarism and a scandalous act.

Complicity Facilitation of unethical behavior in academia and research is termed as complicity.

Computer-aided plagiarism When an author uses artificial intelligence or another similar technology to produce a piece of writing.

Conceptual plagiarism The act of stealing someone else's idea or framework or design without any acknowledgment of the source.

Conflict of interest A situation where personal, financial, or other interests could compromise or appear to compromise an individual's impartiality or professional judgment.

Contract cheating A form of academic dishonesty where students outsource their academic work to third parties, undermining the authenticity of academic assessments.

Copyright The legal safeguard that protects original creators of text or graphics by recognizing their authorship of the products of their intellectual labor. Think of it as a "no trespassing" law for text, graphics, inventions, and ideas.

Cross-cultural plagiarism The practice of taking ideas, art, or knowledge from another culture—often marginalized—without honoring the original creators or their specific cultural contexts.

Declaration of Helsinki A set of ethical guidelines developed by the World Medical Association, focusing on research involving human subjects, particularly the need for informed consent and protection of participants.

Ethics Are the standards that guide our behavior, both as individuals and as members of organizations.

Fabrication A very severe type of academic misconduct in which a researcher/faculty alters the data or invents the information without actual work.

Falsification Involves altering, manipulating, or omitting the existing data or information in order to support false findings or results, e.g., changing methods, materials, research instruments, images, results, etc.

Ghost authorship The unethical practice of excluding individuals who made significant contributions to a research project from the list of authors.

Ghostwriting Getting another person to write your assignment or thesis or synopsis or paper for publication.

Ghostwritten work The result of someone else's labor, but the author does not reveal the name of the original creator. Often, the original creator is paid to remain anonymous, a ghostly presence behind the work.

Good practices Established methods, techniques, or processes that are recognized as effective, efficient, and ethical in a particular field. Good practices in education might include strategies for teaching, learning, assessment, or student engagement.

Honor code A set of principles or rules governing academic integrity and ethical behavior, often emphasizing honesty, fairness, and respect within a learning community. Many institutions implement honor codes to prevent cheating, plagiarism, and other forms of academic misconduct.

Honorary authorship The unethical inclusion of individuals as authors on a research paper when they have not made significant contributions.

Inclusive community A group or environment where individuals from diverse backgrounds feel respected, supported, and valued. Inclusivity focuses on ensuring equal access, participation, and opportunities for all, regardless of race, gender, disability, or other differences.

Institutional Review Boards (IRBs) An administrative body established to protect the rights and welfare of human research subjects recruited to participate in research activities conducted under the auspices of the institution with which it is affiliated.

Intellectual authenticity The commitment to being honest, transparent, and genuine in one's academic or intellectual work, ensuring that one's ideas, research, and conclusions are original and free from plagiarism or dishonesty.

Intentional plagiarism The act of knowingly submitting someone's work as one's own.

Justice The principle of fairness in research, ensuring that the selection of participants is equitable and that no group is unfairly burdened or excluded from the benefits of the research.

Macro-level strategies Institutional-level strategies that create a comprehensive framework for academic integrity across the institution, involving policy frameworks, support services, and awareness campaigns.

Mega-level strategies Strategies that extend beyond the institution to include collaborations with external stakeholders, fostering a shared ethical framework across educational communities and society.

Meso-level strategies Departmental or faculty-level interventions that promote academic integrity through systematic collaboration, policies, training, and syllabus integration.

Micro-level strategies Individual-level strategies focusing on personal actions and development, including education, mentorship, and technology awareness, to encourage ethical behavior.

Mosaic plagiarism When a writer combines or merges several ideas or pieces of writing from various sources without crediting the original authors. The resulting patchwork may look like a good piece of research but is dishonest because it hides the sources and contributes nothing original or substantial.

Non-maleficence The obligation of researchers to avoid causing harm to participants, conducting thorough risk assessments to mitigate any potential negative effects.

Nuremberg Code A set of research ethics principles developed in response to unethical human experimentation during World War II, emphasizing voluntary consent and the importance of avoiding harm.

Paraphrasing plagiarism When an author states someone else's ideas or expressions in his own words and hides the original sources to make them look like his or her own creation.

Peer accountability A system or culture in which individuals are responsible to each other within a group for their performance and behavior. In academic settings, it often refers to students holding one another accountable for participation, performance, or adherence to group rules.

Plagiarism The act of claiming to be the original creator of someone else's words, ideas, sentences, or expressions. This constitutes a major breach of trust in academia and the world of publishing.

Policy for good practice Institutional or departmental policies designed to foster academic integrity by setting standards and expectations for ethical behavior in academic work.

Principal researcher Is someone who leads a research project and responsible of everything and everyone in that project.

Proctoring The act of supervising or monitoring students during an examination to ensure compliance with testing regulations and to prevent cheating. Proctoring can be done in person or through online systems.

Research Ethics Committees (RECs) Another nomenclature for the Institutional Review Boards. This term is vastly used in west European countries.

Research To study a subject in detail, especially in order to discover new information or reach a new understanding.

Respect for persons A core principle in research ethics that emphasizes the autonomy and dignity of participants, ensuring they are treated with respect and provided with the opportunity to make voluntary, informed decisions about their participation.

Retraction The formal removal of a published research paper from the scientific record, usually due to ethical violations or errors in the research process.

Sabotage It is referred to as deliberately preventing another student/faculty from completing their work to gain advantage for self or malign another person (Schmelkin et al., 2008).

Scaffolded assessment A teaching strategy that builds students' skills progressively, reducing opportunities for academic misconduct by providing continuous feedback.

Self-plagiarism When an author recycles his or her previous work without informing the readers that it is a recycled work. This is often done to increase the number of one's intellectual works.

Stakeholders Individuals, groups, or organizations that have an interest or investment in a particular issue, project, or organization. In education, stakeholders often include students, faculty, administrators, parents, and external partners.

Sustainability The practice of meeting current needs without compromising the ability of future generations to meet their needs. It involves maintaining ecological balance, conserving resources, and fostering long-term environmental, social, and economic health.

Twelfth strategy Typically refers to the twelfth point or element in a larger strategic plan. The specific meaning depends on the context of the strategy being discussed, such as an educational, corporate, or institutional plan.

Unintentional plagiarism When an author fails to realize the external sources of his or her inspiration, accidentally fails to cite them, or coincidentally has the same idea as someone else. Whatever the reason, it is still plagiarism—except that in this case the intention to plagiarize does not exist.

Vulnerable populations Groups that may be at higher risk of harm or exploitation in research, including children, people with cognitive impairments, and marginalized communities, requiring additional ethical safeguards.

MIX
Papier aus verantwortungsvollen Quellen
Paper from responsible sources
FSC® C105338
www.fsc.org

If you have any concerns about our products,
you can contact us on
ProductSafety@springernature.com

In case Publisher is established outside the EU,
the EU authorized representative is:
Springer Nature Customer Service Center GmbH
Europaplatz 3, 69115 Heidelberg, Germany

Printed by Libri Plureos GmbH
in Hamburg, Germany